AQA Design an Technology

Product Design

Brian Russell

Krysia Ballance

Andrea Bennett

Nicola Deacon

Jeff Draisey

Series editor

Brian Russell

OXFORD
UNIVERSITY PRESS

OXFORD
UNIVERSITY PRESS

Great Clarendon Street, Oxford, OX2 6DP, United Kingdom

Oxford University Press is a department of the University of Oxford. It furthers the University's objective of excellence in research, scholarship, and education by publishing worldwide. Oxford is a registered trade mark of Oxford University Press in the UK and in certain other countries

British Library Cataloguing in Publication Data
Data available

978-1-4085-0276-1

10 9 8 7

Printed in China

Acknowledgements

Cover: Jim Wileman

Illustrations: Map by Alasdair Bright and technical illustrations by Graham White at NB Illustration

Page make-up: Fakenham Photosetting Ltd

P. 7 Alessi, 1.0A Sipa Press/Rex Features, B and C Philippe Starck, D Design Council; 1.1A Chad Ehlers/Getty Images, C Science Photo Library; 1.2A Ann Ronan Picture Library, B Edimedia Archive, C Corbis, D Vassily Silla, E Public domain, F Public domain; 1.3A Thoresby, B Getty Images, C Ettoire Sottsass, D Alessi; 1.4A Apple Corporation, B Citroen, C Public domain, D Honda, E Apple Corporation, F Alamy. 2.1A i and iii Public domain, ii Science Photo Library/Victor de Schwanberg, iv Nokia, v Research in Motion, vi Apple Corporation, B both Dyson UK, C Toyota (GB) plc, D i Art Directors ii Fotolia; 2.2A Alamy, B LiveAuctioneers LLC, C HMV, D i isStockphoto ii topfoto, E i Ann Ronan Picture Library ii iStockphoto; 2.3B Alamy, C BBC Motion Gallery. 3.1A Reusch, B Diethelm Keller Brands, C and D Nathan Allan Photography; 3.2A and B Bang & Olufsen, C Dyson UK, D Spyder, E www.graphxllc.com, F Alessi, G Alamy, H and I Nathan Allan Photography, J Apple Corporation; 3.3B Jaguar; 3.4A Dyson UK/Getty Images; 3.5A iStockphoto, B Heston Blumenthal, C Rex Features, D and E Ron Arad. 4.1A Nathan Allan Photography. 4.2A Alamy, B Seviroli Foods, C public domain. 4.3A i Heinz Foods ii Innocent Foods iii BMW iv Nike UK v Kellogg's, B Empics. 5.1A MaybachExelero, B iStockphoto, C Jason Newman, D Empics, E iStockphoto; 5.2B Nathan Allan Photography; 5.3A British Standards Institute, B EEC, C and D Public domain, E Sainsbury plc; 5.4Bi Jason Newman, ii iii and iv iStockphoto, D Which? Magazine. 6.1A Public domain, B FSC (Forestry Stewardship Council), C Empics, 6.2C www.carbonfootprint.com, D Fairtrade Foundation, E Nathan Allan Photography. ESQ1(a) First three iStockphoto, last Apple Corporation, (b) ESQ2 iStockphoto, Alamy, Image board author's own: ESQ3 all public domain.
Page 55 all re-use, 7.1A Image Asset Management (hereafter IAM), B iStockphoto, C and D Getty; 7.2A i Goodyear, ii to v iStockphoto, vi Alamy, vii re-use, viii iStockphoto; 7.3B

iStockphoto. 8.2B iStockphoto, C Getty; 8.3A Public domain, B Science Photo Library. 9.1B Alamy; 9.2B Jason Newman, C and D Alamy, 9.3B Nathan Allan Photography, E Public domain, G both iStockphoto; 9.4B iStockphoto, 9.5A both Nathan Allan Photography, C Heidelberg gmbH (Dymatrix cutting station), D to G iStockphoto, H Product code 200024. Axminster Power Tool Centre www.axminster.co.uk, I to K iStockphoto, L Public domain, M iStockphoto, N Alamy, O Jason Newman, P Alamy/fstop, Q iStockphoto; 9.6B and C iStockphoto; 9.8A and B iStockphoto, D Getty Images; 9.10A DK Images, C Alamy/ D Hurst; 9.11A Science Photo Library, C Fotolia; 9.12A Alamy, B iStockphoto; 9.13A Alamy/H. Mark Weidman Photography. P.99 Graham White; 10.1A Alamy, B Public domain, D and both of E iStockphoto, F i Goodyear ii to iv iStockphoto; 10.2A Alamy, B iStockphoto, C Alamy/Scientifica/Visuals Unlimited, D IAM, E iStockphoto, F Public domain; 10.3A Public domain, Bi to iii Nathan Allan Photography iv iStockphoto; 10.4 A FSC (Forestry Stewardship Council), B iStockphoto, D all DK Images; 10.5A and B iStockphoto, C Cedar Fair Entertainment Company, D all iStockphoto; 10.6A, C, D and E all Public domain, F TheGreenCarWebsite.co.uk; 10.7A and B Alamy, C iStockphoto, D lozyska ceramicz; 10.8A and B Nathan Allan Photography, C all Public domain; 10.9A and B DoH; 10.10D iStockphoto; 10.11E and F iStockphoto. 11.1A Public domain, B both iStockphoto, C iStockphoto; 11.2A bighcatering.com.au, B Fotolia/ Thor Jorgen Udvang, D Alamy; 11.3Ai Ojiya Balloon Festival ii and iii Public domain, B US Army, C iStockphoto; 11.4A Prada; 11.5A Public domain, B Alamy, C and D Public domain, E Weidmann, G happydaisydesigns.com; 11.6B and C Public domain, D Nike UK; 11.7Ai Audi ii Skoda Felicia iii VW Golf, Bi and iii Alamy, ii and iv Nathan Allan Photography v Science Photo Library vi Fotolia. ESQ4 i and iii Public domain, ii and v iSockphoto, vi and vii Alamy.
Ch. 12 Header Strip iStockphoto.
13.1C Alamy/ John Warburton-Lee Photography; 13.3B Fairtrade Foundation, Ci Forestry Stewardship Council ii Rainforest Alliance, D i International Association of Natural Textile Industry/www.global-standard.org ii www.uberreview.com. 14.3i and ii iStockphoto, iii Alamy. 15.2B Yuanda, C Alamy. With thanks to Wade Deacon School in Widnes for use of original student projects.

Although we have made every effort to trace and contact all copyright holders before publication this has not been possible in all cases. If notified, the publisher will rectify any errors or omissions at the earliest opportunity.

Links to third party websites are provided by Oxford in good faith and for information only. Oxford disclaims any responsibility for the materials contained in any third party website referenced in this work.

Contents

UNIT ONE

UNIT TWO

The publisher has worked hard to ensure this book and the accompanying online resources offer you excellent support for your GCSE course. You can feel assured that they match the specification for this subject and provide you with useful support throughout your course.

These print and online resources together **unlock blended learning**; this means that the links between the activities in the book and the activities online blend together to maximise your understanding of a topic and help you achieve your potential.

These online resources are available on  which can be accessed via the internet at www.kerboodle.com/live, anytime, anywhere. If your school or college subscribes to kerboodle you will be provided with your own personal login details. Once logged in, access your course and locate the required activity.

For more information and help on how to use kerboodle visit www.kerboodle.com.

How to use this book

Objectives

Look for the list of **Learning Objectives** based on the requirements of this course so you can ensure you are covering the key points.

Study tip

Don't forget to read the **Study Tips** throughout the book as well as answer **Practice Questions**.

Visit schools.enquiries.uk@oup.com for more information.

Practice questions are reproduced by permission of the Assessment and Qualifications Alliance.

Introduction

The book structure

This book is divided into two units which correspond to the units in the AQA Product Design specification.

Unit 1 looks at the following topics which will be tested in the written paper:

- Materials and components
- Design and market influences
- Processes and manufacture

Unit 2 provides guidance on how to be successful with the Controlled Assessment unit.

This book prepares you for the AQA Product Design course. It provides the knowledge you need to understand the subject and you will be able to test yourself with practice questions. You will be carefully led through the demands of the Controlled Assessment task. There are examples of high quality students' work together with a detailed commentary.

Product Design

Our world is full of products that have been designed to meet the needs and wants of different groups of people. From the moment you wake in a morning to the time you return to your bed, your whole day will be influenced and affected by these products. They help to shape your lifestyle and also define you as the person you are. Products are usually made in large quantities with the aim of meeting the needs of different target markets. They must be manufactured to an appropriate quality and presented for sale at a suitable price. To achieve this designers have to make many compromises with regard to styling, materials and manufacturing processes as well as taking account of moral, ethical and sustainable issues. Product Design is complex and it is hoped that by reading this book you will be fully aware of the issues affecting this subject.

Designing

To be a good designer it is important to understand how products have developed over time. You will need to look back and learn about the main design periods over the past century. You will also need to look at the work of famous designers and the design icons they were responsible for.

Designing often involves making small improvements to existing products but occasionally new, radical designs can be produced. You will learn techniques that will help you become a creative designer and learn about different methods of presenting your ideas to others.

Controlled Assessment

The Controlled Assessment tasks in this book are designed to help you prepare for the tasks your teacher will give you. The tasks in this book are not designed to test you formally and you cannot use them as your own Controlled Assessment tasks for AQA. Your teacher will not be able to give you as much help with your tasks for AQA as we have given with the tasks in this book.

A designer must consider the impact their product is likely to have on others. In this country we consume far more of the world's resources than we should and many of the products we buy exploit people in other countries. By reading this book it is hoped that you will not only become a better designer but a more informed consumer.

◼ Making

If you are to develop your design into a working product you need to know about materials. You will learn about the advantages and disadvantages of using a variety of materials. You need to know about different methods of manufacture. You will learn how to cut, shape, form, mould, condition, assemble and finish a range of materials. You will also learn how to work safely and how to use industrial methods of manufacture to improve the accuracy and consistency of the products you design.

◼ ... and finally

Product Design is an exciting and very rewarding course. It will involve you in a great deal of decision making and hard work. You will need to plan ahead and become very organised. In the end, you should finish up with a wide range of knowledge skills and understanding that will be useful to you over the coming years, and hopefully, you will have designed and made at least one product that you can be really proud of.

Objectives

Understand what product design means and how this affects the role of a designer.

Design and market influences

■ What is product design?

Everywhere you look you are surrounded by product design. From the moment you wake, your senses are bombarded by the sounds, colours, smells, tastes and textures of products; think of the design of your alarm clock, your bedroom, even your breakfast! The design process has been applied to things we use every day to make our lives easier, from the built environment we live and work in, to smaller products such as our mobile phones.

The product designer

Most products have been designed in order to provide solutions to specific problems identified by individual designers, clients or corporate organisations. Some are totally new ideas, but most are the result of improvements made to existing products.

The product **designer** must respond to a host of issues relating to the consumer for a design to be successful. Sometimes products are designed to meet our basic human needs (food, water, shelter) or for the specific needs of a group of people, such as the design of wheelchairs for the disabled. Often in our society, products are designed to meet our desires rather than our needs, and the product designer must be guided by style and fashion, people's likes and dislikes, and cost.

Profile of a product designer

Now let us look at a famous product designer – Philippe Starck, one of the best-known **contemporary** designers in the world. He is an accomplished architect and has received public acclaim for his interior building designs. He has also proved to be an innovative product designer who has been able to move with the times and remain in the public eye.

Starck's designs in the 1980s and 1990s were influenced by style and fashion, but in the 21st century he has begun to promote the ethos that honesty and integrity should be the main principle of design. His belief that products should be durable and long lasting, and should not be created as 'throw away artefacts' was ahead of its time, and it was this foresight that led to his continued success.

A *Philippe Starck*

B *Icon of industrial design: the Juicy Salif designed by Starck for Alessi in 1990*

C *Philippe Starck's personal wind turbine: 'Democratic Ecology'*

∞links

Dick Powell: Introducing Product Design www.designcouncil.org.uk

Key terms

Designer: a producer of designs which fulfil a need or fashion trend.

Contemporary: belonging to the present day.

Consultancy: an agency that provides professional advice.

Podcast: audio or video media file distributed over the internet – named from the words 'iPod' and 'broadcast'.

Design consultancy

It is becoming more common for designers to work within a team of design consultants. Seymourpowell is a design **consultancy** founded in 1984 by Richard Seymour and Dick Powell, and is now part of the Lowey Group. The company offer their customers a range of product design skills including market research, branding strategies, product design and development, and packaging solutions.

Design Council

The Design Council is the national body for design offering information online about product design and the UK design industry. Designers such as Dick Powell write many of the articles and information is available for teachers and students. **Podcasts** featuring audio and visual presentations by experts in the field of Product Design can also be downloaded free through the Design Council website.

D *The Design Council's logo*

Activity

Max Fordham is the 2008 winner of the Prince Philip Designers Prize, awarded annually in recognition of outstanding lifetime achievement in design and presented by H.R.H. the Duke of Edinburgh at a special ceremony at the Design Council. Find out what he designed and why his designs achieved this honour.

Summary

Product design involves products being produced to meet human needs and wants.

Product designers design products to solve a problem or meet a specific need.

Design consultancy involves a group of people solving a range of design-related issues for a client.

The Design Council is a national group that offers information relating to product design.

1 The evolution of product design

1.1 Why products change over time

Our modern world is the culmination of years of development and innovation. A number of important factors help to move products forward, including improvements in the way they are manufactured, the selection of materials the designer has to choose from, and the needs of an ever changing society.

Objective

Be able to identify key reasons why products have evolved over time.

■ Materials and manufacturing

Developments in new materials have allowed designers to develop products that improve our standard of living. Before the Industrial Revolution, available materials were limited to those that could be manufactured by hand from local resources.

With the development of the steam engine, iron and steel production led to **mass-produced** goods. Textiles could be shipped from abroad and cloth produced in quantity. Building, transport and the production of energy changed the face of our towns and countryside.

The development of synthetic plastics from coal, oil and gas brought us another revolution in manufacturing and by the start of the Second World War **Bakelite** was widely used for everyday household products. Mass production in modern plastics gave a wide range of goods to everyone, at an affordable price.

New manufacturing ideas such as just-in-time (JIT) and the widespread introduction of computer-aided design (CAD) and computer-aided manufacturing (CAM) have improved product quality, yet reduced costs to the consumer.

Key terms

Mass-produced: made in great quantity by a standardised process.

Bakelite: a synthetic plastic named after its inventor – L.H. Baekeland.

Utility: describes an item made for its usefulness only.

A *An automated production line*

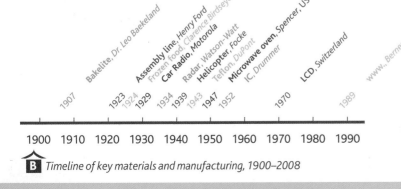

Bakelite, Dr. Leo Baekeland — 1907
Assembly line, Henry Ford — 1923
Frozen food, Clarence Birdseye — 1924
Car Radio, Motorola — 1929
Radar, Watson-Watt — 1934
Helicopter, Focke — 1939
Teflon, DuPont — 1943
Microwave oven, Spencer, US — 1947
IC, Drummer — 1952
LCD, Switzerland — 1970
www., Berners-Lee — 1989

1900 1910 1920 1930 1940 1950 1960 1970 1980 1990

B *Timeline of key materials and manufacturing, 1900–2008*

∞links

See 7.2 for more on just-in-time production.

See 8.1 for more information on computer-aided design and manufacturing.

Social factors

People's views on what is acceptable or fashionable have changed through the ages. Wearing a bikini in England, for example, would once have been considered outrageous, and in some countries bare legs are still unacceptable. Products must therefore be designed to appeal to consumers. In a world led by high street fashion, products are seasonal and designers must constantly change colours and styles.

Political changes, especially regarding the economy and during times of war, also affect the way products are developed. During the Second World War, efforts were concentrated on arms production; many other goods, including furniture and clothes, were limited to cost-effective **utility** items.

Environment

It is only recently that we have begun to understand the impact on the environment of the consumer society. The 'disposable' culture is gradually being replaced with a drive to rethink, refuse, reduce, re-use, repair and recycle. In some cases consumers are willing to accept a change in quality and perhaps even higher prices for 'green' products such as organic food. Consumers are choosing to refuse to accept products that have a negative impact on the environment. Many governments are now introducing specific laws to reduce the carbon footprint of certain products. The motor industry is working hard to produce cleaner cars for the future.

C *Mazda MX5 converted to part solar power*

Activity

Talk to an elderly relative or friend and ask them about the products they remember from their childhood. Are these products still available? How have they changed over the years?

Study tip

It's a good idea if you can describe how products have evolved over time and what factors have influenced this evolution. It helps to learn the names of key materials such as Bakelite.

Summary

Developments in technology have affected materials and manufacturing. New materials have been produced, and the way we make products has changed.

Social factors have an impact on the products we use.

Products for a 'greener' future are being developed.

1.2 Design movements

In 15th-century Germany, Johannes Gutenberg invented a printing press to replace the hand-carved wooden blocks traditionally used for printing text. This was one of the first examples of machine manufacturing. Most products were still made in the home or in workshops until power-driven machinery allowed them to be made in quantity.

The Crystal Palace designed by Joseph Paxton for the Great Exhibition of 1851 was a triumph of design and engineering. The Great Exhibition helped show that many products were too elaborate, of poor quality and difficult to operate. By 1900, manufacturers had begun to understand that products needed to reflect the needs and wants of consumers. The role of the product designer was born.

A *The Crystal Palace was a masterpiece of design and engineering*

The various design styles that exist show a range of colours, shapes, textures and forms and were influenced by social and political circumstances across the world.

The Arts and Crafts Movement

William Morris founded the Arts and Crafts Movement at the end of the 19th century. His designs for wallpaper, furniture and textiles were inspired by organic shapes and patterns found in nature. Morris was a socialist, firmly against poor working conditions and the damage done to the environment by industrialisation. He was keen to promote the production of quality products. His work used expensive materials and traditional techniques that only the wealthy could afford.

B *Textile design by William Morris*

Art Nouveau

This design style took its name from a shop that opened in Paris in 1895. Based on the organic lines of climbing plants and Japanese art, it was popular with designers of glass, furniture, fabrics and wrought ironwork. Some of the most famous designs are the lamps of Louis C. Tiffany, and the work of René Lalique.

Modernism

Modernist designs were made ergonomically, using appropriate materials and very little decoration. Designers like Charles Rennie Mackintosh moved away from organic lines and started to use geometric shapes, which were easier to mass produce.

Bauhaus

The Bauhaus was a school of art and design, founded in Germany by Walter Gropius. Between 1919 and 1933, Bauhaus designers used modern materials and mass-production methods. Experimental work using colour and form was encouraged to produce designs that were both artistic and skilled, while following the underlying principle that **form** should follow **function**.

Art Deco

This fashionable and glamorous period of design was influenced by other design movements, as well as ancient Egyptian art. Used in interior design between 1920 and 1939, its influence can also be seen in the architecture of the time. Clarice Cliff, a famous designer of ceramics, decorated her work using this bright, bold style until the Second World War when it became illegal to use time and resources on decorating products.

De Stijl

Using basic shapes and primary colours, this movement took geometric design to another level. Founded in Holland by a group of painters and architects, including Theo van Doesburg, it was the inspiration for a range of furniture and architecture that used only the essential form and colour in the design.

C *Tiffany lamps are a shining example of Art Nouveau style*

D *New materials were used for furniture*

E *A vase by Clarice Cliff*

F *De Stijl furniture by Gerrit Rietveld*

Summary

Art movements and the artists behind them have continually moved painting, sculpture and design forward by uniting around shared visions and philosophies.

The birth of the designer label

In the 1930s there was great interest in the 'need for speed', and competition to be the fastest on land, sea and in the air. This led to the inclusion of 'streamlining' in everyday objects. The national electricity grid was growing and electrically powered household items, such as radios and vacuum cleaners, were becoming popular and their design reflected the new streamline style.

The growth of industry and production development during the Second World War provided a new generation of skilled engineers and designers. Technological advances made during the war, in materials, electronics and ergonomics, were later used by professional designers in the design of everyday products.

A Ford Zephyr: an early covetable label

The growth of youth culture

Post-war reconstruction meant that rationing continued into the 1950s – by which time people were looking for new exciting products. Youth culture had a huge impact on design and fashion. In the 1960s young people fought against traditional lifestyles and behaviour. Mary Quant's design for the mini-skirt, for example, was hugely popular. Designers and artists were influenced by the British and American music industry and the hippie movement. Futuristic designs were a feature of the 1970s, with a growing demand for automated products; these were also featured in popular films and television programmes.

Post-modernism

Designers, architects and artists who did not like the simple modernist designs used a variety of decorative finishes to make their products more aesthetically appealing. Post-modern designs were popular among the young, who drew upon a range of styles for inspiration, led fashion and encouraged the development of a growing consumer market. The Memphis Group, a 1980s group of designers from Italy, produced designs typical of post-modernism. Designs such as the Carlton dresser by Ettore Sottsass (**C**), used materials and colour to provide visual impact. Functionality was of minimal importance.

B Mary Quant (centre) and friends showing off the mini-skirt

C *Ettoire Sottsass's Carlton dresser*

∞links

For information on:

market pull, see Chapter 2

the power of branding, see 4.3.

Study tip

The term 'brand image' is an important one. Make sure you understand what it means. It will help if you can also write about commercial influences on product design.

Key terms

Brand image: this identifies the company who made the product and gives a particular impression of its qualities.

Designer labels

The 1990s saw a change in the way people bought products. The designs themselves became less significant, with the emphasis being on the designer instead. This new marketing age brought a demand for 'designer' products – celebrity image, promotion and packaging became all-important.

Who are the designers?

Brand image is now a major factor in the sale of goods, and companies work hard to be at the top in their product area. Many individual designers are now hidden behind the brand label. Alessi, for example, has a strong brand image supported by a team of designers, and produces high-quality household goods.

D *Alessi Anna G corkscrew*

Summary

New production technologies allowed products to be 'streamlined'.

Post-war, young people became consumers, and started looking for products that portrayed their individuality.

In post-modernism, designers used colour and style to great effect but did not dwell on the function of the product.

As companies compete to be market leaders, brand image becomes more important.

Nothing has driven forward product design as much as the advent of new technologies. In the last 20 years personal communication and entertainment have been transformed with the development of the mobile phone and an explosion of media devices.

Blobjects

Product design using CAD (computer-aided design) and CAM (computer-aided manufacturing) reduces styling constraints and has brought about the **blobject**: a product displaying organic, colourful, free-flowing shapes. Designing on a computer allows us to create shapes and designs that could not be done easily by hand. Products can then be manufactured using injection-moulded plastic and cast or pressed metal. Examples of this can be seen in the Apple iMac computer and the Citroën C1 and C3 cars. Buildings designed in this way, such as Norman Foster's 'Gherkin' tower in London and the Sage Gateshead music centre in Gateshead have been termed 'Blobitecture'.

A *An iMac*

B *A Citroën C1*

C *The 'Gherkin' in London*

Anthropomorphism

Giving human characteristics to inanimate objects is called **anthropomorphism**. The **human interface** is 'softer' and makes the product more aesthetically pleasing. Everyday objects such as remote controls use this technique by placing the buttons to form the features of the human face. The humanoid robot ASIMO, designed by Honda, is a fantastic example of anthropomorphism and 'intelligent' technology.

Gizmos

Microelectronics and the miniaturisation of components have revolutionised the development of multi-functional electronic devices such as the mobile phone and personal digital assistant (PDA). **Gizmos**, as they are collectively known, have a short life as companies competing for the market share add more and more features. Adding more features certainly adds value but also serves to make products more complex.

D *ASIMO*

nano-chromatic

E *Evolution in design: the iPod nano*

New and smart materials

Smart materials respond to their environment and allow us to personalise products. When included in a product they often subtly improve its performance – without increasing its complexity. For example, smart grease gives the closing mechanism of a CD player a beautifully smooth action. Many new materials allow existing products to be miniaturised or give us the ability to dramatically improve their performance. Optical fibres can transport light over 200 km and are used in telecommunications, computers and in surgery.

F *New developments in fibre optics are changing the way technology can be used*

links

For more information about smart materials and new materials, take a look at 11.4 and 11.5 on pages 128–131.

Remember

Most products use a range of materials in their manufacture. Smart and new materials can be used to improve parts of the overall product.

Key terms

Blobject: a product designed using CAD or CAM to reduce styling constraints.

Anthropomorphism: using human features on objects to improve the human interface.

Human interface: the relationship between the product and the user.

Gizmo: a small, multi-functional device.

Activity

1 a Working in small groups, describe to each other the advantages and disadvantages of adding features to an electrical communications device.

 b What do you think products will be like at the end of this century? Explain how developments in new technologies might further change product design.

Summary

New technology has driven design forward.

Anthropomorphism uses human features on objects to improve the human interface.

Microtechnology and miniaturisation have revolutionised products such as mobile phones.

Smart materials respond to their environment, and can reduce the size and complexity of products.

links

Find out more at:
www.apple.com/itunes

For more information on the development of a new product, see:
www.theapplemuseum.com

2 Meeting consumer needs

2.1 Technology push and market pull

New products are continually being developed to meet the needs and desires of the consumer. Over the past hundred years continuous technological development has had a significant impact on the way we live and consumer expectation has escalated.

Technology push

Technology push is a process in which the new advances in science and technology, discovered in research and development departments, are passed onto a design and development team so that the advances can be used innovatively in technology and in the making of products. Improvements have been made to a range of existing products using new technology. For example, mobile phones have developed from a simple communication device for making calls into complex 'personal digital assistants' (PDAs).

A From carphones to PDAs

New products have been produced using new technologies. After the Second World War experiments using radar resulted in the creation of the first laser (light amplification by stimulated emission of radiation). At the time it was thought that the laser would be useful for further scientific research, but currently it is applied to a range of products such as CD and DVD players, widely advertised eye surgery, and the CAM machines often used in school workshops.

Market pull

Market pull describes how the inspiration for new products often comes from the needs of society. Market research and analysis of existing products can help manufacturers to improve their products so that they can:

- sell to new customers to increase the company's market share

B Dyson have responded to market pull

- encourage brand loyalty
- persuade customers to choose their brand and product over alternative brands and products.

A good example of market pull, where customer demand is high, is the need for smaller, 'greener' cars. Many vehicle manufacturers now use existing technologies to provide high specification small cars with better fuel economy and less environmental impact.

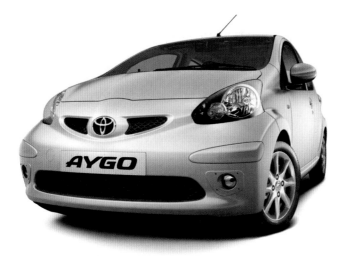

 Rising to the challenge: to sell well, 'greener' cars must also have appealing design

Obsolescence

Some companies deliberately plan to minimise the 'life' of a product in order to maintain sales of future products. This is called 'built-in' or 'planned' obsolescence. Products such as single-use cameras and ballpoint pens are designed to last only a short time and are often made from non-renewable materials.

Obsolescence can also be due to changes in fashion and the addition of alternative features on products. Despite the fact that mobile phones can continue to work well for many years they often become **obsolete** after about 12 to 18 months due to changes in phone styling and the range of features available.

D *Products with built-in obsolescence*

∞links

For examples of technology push, check out the following websites:

www.dyson.co.uk
www.phonehistory.co.uk

Activity

1. Using the internet or magazines, find examples of a wide range of products. Divide them into two groups:

   a those which have been developed through the use of new technologies, and

   b those which have been produced due to the wants of the consumer.

Discuss your findings.

Summary

'Technology push' refers to new scientific and technological advances leading to improvements in a range of existing products.

'Market pull' refers to consumer demand providing the impetus for the development of new products.

Obsolescence is a feature of many manufactured goods. Products become outdated or are designed to wear out after a certain period of time.

2.2   Music on the move

Many changes and developments in product design are due to technological advancements, new materials and automated production processes. Social developments for example, women going out to work and fashion also have a huge impact.

Music has been popular entertainment throughout the ages. By the 20th century it had become possible to reproduce music mechanically. This is a good example of a product continuing to evolve through to the 21st century.

Cylinder music box and disc musical box

The cylinder music box, invented in Switzerland, was an important type of **mechanical** music a century ago. After winding, the music is played on a 'comb' of teeth which is struck by the pins on the cylinder. Each musical box offered a selection of tunes that played fairly quietly, amplified by the table it was placed on and by closing the lid.

Musical boxes using interchangeable discs, such as the Polyphon, became popular as they were less expensive, louder and offered more choice of tunes. The upright model proved popular for the commercial market, to be used in public houses, and was often coin-operated.

A *A cylinder musical box*

B *A Polyphon disc player*

Phonograph and gramophone

The phonograph revolutionised music technology. Previously, the music was generated by the musical box itself, whereas it now became possible to record voices and instruments in grooves on a wax cylinder. The wind-up gramophone, popularised by His Masters Voice (HMV) triumphed in the marketplace. Music was recorded on discs or 'records' made from shellac (a resin material made from the lac bug), which were brittle. The grooves had to be sufficiently far apart to avoid collapse. Each record could only hold one tune.

Record player

The **electrical** record player followed, allowing discs to be changed without the need for winding. Materials technology advanced the record itself. Vinyl discs allowed the grooves to be much closer and

Objectives

Reinforce understanding about the way products evolve and the effect of continuous improvement on this evolution.

Understand how continuous improvement can help manufacturers increase their market share.

Key terms

Mechanical: working via a mechanism without direct human intervention.

Electrical: operated using electricity.

Laser: amplification of an output of light producing an intense beam.

Digital music: analogue music is transferred into a computer data file.

links

Listen to cylinder and disc music boxes: www.mbsgb.org.uk

For more information on technological push, see 2.1

For more information on mechanical control components, see 10.11

C *The wind-up gramophone (HMV) was later available in a portable version which also held a selection of records*

each long-playing record (LP) could hold several tunes, and a 'single' allowed two (one on each side of the disk).

Walkman, Discman and MiniDisc

In the 1980s, music was recorded on to tapes that could be played on the innovative 'Sony Walkman', allowing truly portable music for the first time. The 1990s follow-up was the Sony Discman, with electronically recorded CDs using **laser** technology. The MiniDisc, introduced in 1992, was of similar CD quality. It was popular in Japan, but did not really catch on in Europe.

MP3 and iPod

Digital music technology has brought music to its present format. We can now download and store music on MP3 players and the fashionable iPod, which are small enough to be kept in a pocket. They can hold a large number of tunes on either a hard drive or flash memory, and many can also be used as external data storage devices.

The future

Over the years the market has demanded more portable devices. We want products to do more than just play music: for example the iPhone and other PDA (personal digital assistant) phones allow us to carry one product for a range of uses. They can be integrated into our vehicles or clothing, and we can download podcasts and watch TV via the internet on them.

We have looked at the development of music over the last hundred years. Consider what we might be listening to by the end of the 21st century.

D *Electrical record players took away the need for winding*

E *The iPhone, which fulfils many different functions, is replacing earlier machines such as the Sony Discman*

Consumer issues and awareness

Products evolve for a variety of reasons, including technological push and market pull, social changes, and fashion. Products also evolve because manufacturers are continually trying to better their products, so they spend time assessing their current work.

Continuous improvement

Companies that work to recognised standards such as BS EN ISO 9000:2000 can demonstrate production control and product quality to their customers. Many internationally recognised companies, such as Toyota and Rolls-Royce, involve employees at every level in the process of **continuous improvement** (CI). Groups of employees, known as quality circles, ensure the product is continually improving through contact with the client and updating the product specification. A range of issues may be taken into account including:

- new technology
- legislation
- improvements to production methods
- fashion
- impact upon the environment
- product maintenance
- product durability or obsolescence
- costs
- feedback from the client or consumer.

This approach is called Total Quality Management (TQM).

Planning

Communication

Culture

Peformance

People

Process

Commitment

A Total Quality Management

Market research

In order to make sure that the product meets the requirements of the client and consumer, companies often carry out **market** research. Specific groups may be targeted in order to get a clear understanding of people's needs and wants. A variety of methods may be used, including:

- questionnaires – often sent out through the post, and cover a wide socioeconomic area
- surveys – often carried out locally, targeting specific consumers
- testing – consumers are given samples of goods to test and feedback forms to complete
- sales – the number of goods sold gives an indication of the product's popularity
- longitudinal studies – observing how consumer tastes change over time.

B Market researchers are paid to find out what we think

Product analysis

Product designers rarely design a product that is entirely new. Most ideas are based upon influences such as nature, artists or designers, design movements or similar products. Product **analysis** is a key feature in researching and developing a new design and in CI. Analysing existing products can help us to keep and use ideas that work well and disregard those that are less effective.

We rarely get the opportunity to test a range of products ourselves, but there is usually a team of experts willing to take on this task for us and pass the information on via television or the internet. *Which?* (published by the Consumers Association), analyses products and prints the reports both in its magazine and to subscribers on the internet. Comparison websites also analyse a range of products and services, allowing us to make judgements about features and quality before making a purchase. The BBC television programme *Top Gear* was first screened in 1977 as a conventional motoring magazine show which aimed to road test a range of vehicles. It has since become popular for providing reviews in a humorous style with some outlandish tests that are beyond the reach of most people. The *Top Gear* 'Cool Wall' was produced to compare the style of cars as well as their specification. This idea could be applied to the form and function of almost any product, putting the best 'cool' ideas at one end, running through to the worst 'un-cool' ideas at the other.

A light hearted approach to product analysis is taken by the Hat Trick Productions programme for BBC, *Room 101*, where celebrities are invited to nominate products, services or even people to be consigned to Room 101. The title refers to the location in George Orwell's novel *Nineteen Eighty-Four* within which, for each person, is the worst fear they can imagine. Some of those items to have been nominated include: education, mobile phones, and the small piece of cotton that holds a new pair of socks together!

Study tip

Look at a selection of *Which?* reports to help you think of different products to write about.

Key terms

Continuous improvement: making designs better.

Market: the target group a product is aimed at.

Analysis: discovering the important features of the design problem.

∞links

Check out the design website:
www.bsieducation.org

Look up the *Top Gear* take on 'cool' and 'uncool' by following the 'Top Gear Cool Wall' link at
www.jeremyclarkson.co.uk

C *The Cool Wall: a new twist on market research*

Activities

1 Produce a *Top Gear* style 'Cool Wall' for a product of your choice.

2 Nominate a product – that you think has significant faults – to be consigned to Room 101. Devise a rational argument against the product to put to the rest of your class and teacher. Do they agree? If not, consider their reasons.

Summary

Increasingly, companies use recognised standards to ensure the quality of their products and production methods. This is known as continuous improvement (CI).

Market research enables designers to find out information from the users of products to help improve designs.

Product analysis allows designers to look at existing solutions to find out which ideas already work well and why.

3.1 Developing a product for an identified need

Designers do not design just for the sake of designing. Innovative products are always sought after, be they new food products, new textiles products or a product using resistant materials.

Self-heating gloves

Designers often conduct research to identify products which can be improved. Such research can show that consumers need or will want something better than is currently available. As many winter sports enthusiasts may find themselves out on the ski slopes for long periods, designers of 'interactive wear' felt there was a niche market for a self-heating ski glove that would keep people warm for longer. So their **innovation** was a state of the art ski glove, which is lightweight and self-heating. The product is particularly useful for people with poor circulation in their hands, but should be enjoyed by many others too.

Highly specialised conductors have been attached inside the fabric and efficient lightweight lithium ion batteries supply power for heat. These batteries can be distributed on or in the garment to give heat for several hours of skiing.

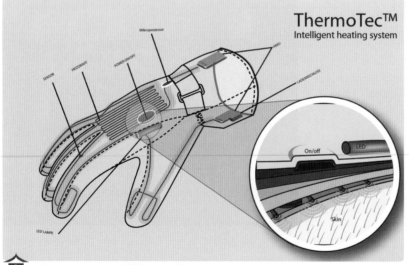

ThermoTec™
Intelligent heating system

On/off   LED

Skin

A *An interactive wear self-heating ski glove*

⚮links

For more information about technological push and market pull, see 2.1 and 2.2.

Baby feeding products

Designers are always looking for new and innovative ideas to help develop the range of feeding products for mothers and babies. Tommy Tippee products were developed to help parents know when their baby's food is at right temperature so it will not harm the baby's mouth. The bowls and spoons are made from smart materials which change colour: bright yellow when very hot and pink when the food is the correct temperature.

One Touch can opener

One Touch is a company of clever designers who have developed a unique, hands-free can opener aimed at changing the lives of people with weak or arthritic hands. Seven million products have been sold worldwide and the company has also won five major international awards for design and quality.

B *The One Touch can opener*

C *Smart Tommy Tippee spoon for babies*

Case study

Technology just blows you away

Designers have invented the Rocket Fuel Espresso Coffee Powder Shot, a self-heating coffee drink that contains three ingredients: the coffee drink, sugar and natural guarana. The pack is shaken for forty seconds and heat builds up in the consumer's hand. Designers have targeted consumers who are regularly on the go, and might want coffee at inconvenient times or in inconvenient places, such as on top of Scafell Pike or in the middle of Lake Windermere. Rocket Fuel Espresso will give the consumer a caffeine pick up, but without the need for a flask.

The product uses an exothermic chemical reaction, something that happens when two chemicals are combined to give out heat – a good example of technology push.

D *Rocket Fuel Espresso Coffee*

Activity

1   On these pages, there are four products designed and manufactured for a particular identified need.
   a   Write a design brief for each product.
   b   Brainstorm an analysis for each product.
   c   List the forms of research that you would consider for each of the products.
   d   Have a try at writing a specification for each of the products.

○○links

For more information about smart materials, see 11.4.

For more information on the One Touch Can Opener, see www.daka.com.hk and click on 'small kitchen appliances'.

For more information on the Rocket Fuel Espresso Coffee, see www.rocketfuel.uk.com

Summary

Innovation in product design can harness technology to improve what is already available.

Technology push inspires the design of products with new safety features, and uses science in innovative ways.

Cleverly designed gadgets can make a real difference to quality of life.

Every new product must have a unique reason to exist

Designers surprise and inspire, and create icons for tomorrow. However, not all products are innovative or iconic. Some products are developed to grasp a percentage of market share, or to reduce manufacturing costs.

When designers and engineers developed the successful Bang and Olufsen 'BeoLab 8000' speakers, for example, they combined style with function. The **design engineer** team spent time working out how the organpipe-shaped speaker would stay upright but at such an angle to the flat base. Engineers spent time debating and developing **prototypes** to make the speaker a reality. The engineers considered the weight then designed a pin strong enough to support the base. At the same time, Bang and Olufsen would have been concerned about the economics of the new design. For example: Would it appeal to the market and so make the investment in **product development** worthwhile? How much of their existing manufacturing plant could be used without expensive adaptations?

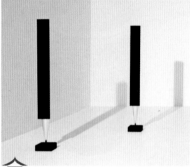
A Beolab 8000 speakers

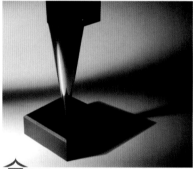
B Supporting pin and base of the Beolab 8000 speaker

C Dyson ball barrow

D Spyder ski jacket

E Ski boot

⬭⬭ **links**

For more information on Bang and Olufsen speaker designs, see:

www.bang-olufsen.com, and

www.beoworld.org

F *Alessi lemon squeezer*

G *Smart car*

H *'Diabetic' chocolate bar*

I *Fat-free and sugar-free yoghurts*

J *iPod Nano*

Study tip

Make sure you can explain what considerations a company must take into account when developing a new design, including, for example, the cost of development and manufacture and earlier versions on which they can build. Always consider what makes a design good.

Activity

1 a Discuss each of the designer products on this page, listing the features from a style point of view.

   b Discuss as a group what kind of engineering challenges would have to be taken into account.

The iPod nano

The iPod nano is small and extremely portable. Apple has successfully created a trend with iPod design and everyone wants one. Designers thought about their target end-users, who were people who liked jogging or walking, and created a small portable device with an easy-to-operate single hand control. They also considered that day-to-day users might like to take their nano to work in their pockets. The outside casing for the nano is made from a polycarbonate plastic material to withstand rough handling, and the designers produced a remote control to allow the nano to be operated from a distance.

Designers considered human factors to help make it to look simple and easy to use, and created a single button control with many functions, including menu navigation and volume control. They also wanted it to resemble the older, hugely successful iPod.

The iPod nano allows users to have their own music collection with them at all times. The simplicity of the design is important – the user believes it has to be easy to use because of its simplistic design. The single-click control wheel, straightforward interface and simple aesthetic design have established the iPod range of products as design icons of their era.

Summary

A successful new product must capture the imagination of potential buyers, and be seen to be better than its predecessors.

Manufacturers need to consider the cost of new developments, and how they can use existing infrastructure.

When style and function meet, a product can become iconic.

Modelling a design proposal for the Jaguar XK8

The Jaguar XK8

The Jaguar XK8 is a classic example of an iconic design. It combines function, comfort and style in a way which captures the imagination; it is no surprise that the XK8 features in a Bond movie.

Ergonomics, safety, efficiency and aesthetics were all essential to the development of this car.

Ergonomics

This covers the way in which the person relates to the car. It includes the amount of leg room; and the position of the gear lever and the dashboard so that they are within easy reach of the field of vision, and so that getting in and out of the car is easy.

Safety

This is perhaps the most important aspect of any vehicle, luxury or otherwise. How well protected will the driver and passengers be in a crash? Are the protective crumple zones sufficiently strong, and are they well positioned? How does the car perform in different road conditions (wet, icy, overcrowded, etc.)?

Efficiency

An important development in the XK8 was the styling of its aerodynamic shape. This is not only aesthetically appealing, but reduces wind resistance thereby increasing the potential speed and fuel efficiency of the car.

Aesthetics

The beauty of the XK8 is what sets it apart from other cars and gives it its iconic status. As well as the sleek exterior, the interior is luxurious, with expensive materials such as leather, to give drivers and passengers the sense of travelling in film-star style.

Modelling the Jaguar XK8

The preparation and the development of this luxury vehicle were lengthy and painstaking. The development work used to present the ideas to the client included:

- marker pen 2D presentations
- a full-size clay model of the car. This was constructed with the help of CNC (computer numerical control) technology. The model was first designed on a 3D piece of software, so that the information on the screen could be realised using CNC machinery. Examples of this sort of technique used in your school might involve a CNC router controlled by CAD (computer-aided design)

Objectives

Consider the factors involved in the design of a product that is to be produced and manufactured in quantity.

Be able to respond creatively to briefs, developing proposals and producing **specifications** for **products** and associated services.

Study tip

Remember that 'modelling' means testing your design in 3D not, modelling on the catwalk.

Make sure you are able to describe how you have modelled a design you have made at school as part of the development stage of your project, including the materials you used.

Key terms

Ergonomics: the study of size, comfort and safety in relation to the human and the product.

Product specification: detailed description of what the product is to be.

Modelling: a way of developing part or all of a 3D product using card, clay, foam, wood or CAD.

∞**links**

For more information about human factors, see 5.1.

Check out the manufacturer's website for more information: www.jaguar.co.uk

- a full-size model. This was created to simulate the finished XK8. After it was designed on screen using CAD software, the 3D model was created using CAM (computer-assisted **modelling**)
- a life-size model of the shell of the XK8. This was constructed to determine the exact fit of all the interior components.

Ergonomic design

Marker presentation

Siting of electronic components

Crash testing

Weather testing

Protoype model

Clay model

Global production

A The XK8 underwent these stages of modelling, and was put through these testing procedures, typical for new car development. See www.jaguar.co.uk.

Additional testing

- **Ergonomics**: the ease with which the driver and passenger could relate to the XK8, including comfort, field of vision and accessibility of electronic gadgets, were all tested by computer simulation and using a full size model.
- **Safety**: a crash test was simulated by computers in order to look at the effect of forces applied to the body of the car during a collision. Full-size prototypes were used to test how the car would handle the road in wet conditions.
- **Efficiency**: computer modelling was used to test this, as were full-size models at a more advanced stage.

Once Jaguar was happy with all aspects of modelling and testing, the XK8 went into production and was sold across the world.

B The Jaguar XK8

Summary

Many factors need to be considered while an idea is in development. These may include ergonomics, efficiency, safety and aesthetics.

Modelling is an important stage in developing a concept and helps the designer share ideas with the client.

Later models can be tested for safety and function using sophisticated computer software.

3.4 Three ways of approaching the design process

Empirical design

By analysing and observing, designers learn and develop fresh innovations. Different designers use different approaches when they begin to design and develop their ideas.

James Dyson used **empirical design** when developing his bagless cyclone cleaners. Empirical design means designing with a trial and error, experimental approach, and looking at past practices and predicting the product's future. Dyson's design process was based on empirical testing, making one change at a time to ensure that each part of his product was as good as it could be. The Dyson Cyclone was the first breakthrough in vacuum cleaner technology since 1901. He made over 5,000 models and prototypes before the DC01 went into production.

Intuitive design

Some designers can make decisions quickly, immediately perceiving what is going to work. People with **intuitive design** skills know their specialist areas very well. You will find intuitive designers in advertising media, magazines and newspapers. Fashion designers are intuitive in predicting fashion trends and food specialists know just what the public wants (although some food technologists, such as Heston Blumenthal, also design empirically).

Systematic design

To help you complete your GCSE coursework, your teacher will encourage you to use a **systematic design** approach to help you meet the assessment criteria. This means that you must break down all the different areas of the design process, such as research analysis, design specification, generation of ideas and development.

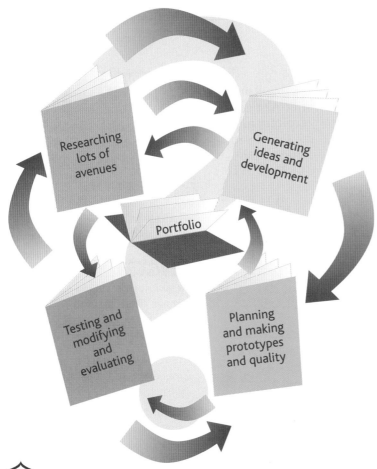

B *The design process involves interaction between any starting points on this diagram*

As designing is not a linear process, you can start your design process anywhere. For example, you could start by analysing an existing product and considering how you might improve on it. Card-modelling instead of sketching, then recording this with a digital camera could be your empirical approach to designing or it could involve researching a particular designer or photographing a particular topic such as nature or cultural influences. If you are designing in this way, you will need to be able to move backwards and forwards in your folder to revise your design thinking.

Source **B** will help you understand the four components that make up your folder.

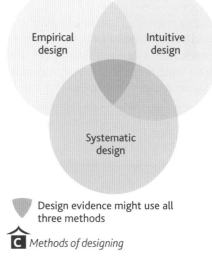

Design evidence might use all three methods

C *Methods of designing*

Summary

Empirical design involves observing and testing each stage of the process.

Intuitive design involves a deep knowledge of the field of work and an ability to predict trends.

Systematic design involves analysis and observation, being prepared to try different methods, and having an open mind.

3.5 Designer case studies

Creativity and innovation can be developed in different ways and there are many starting points for designing and making. A first-hand reference could be photographing **natural forms** and **patterns** as they occur outdoors. **Structures**, such as buildings, and the **geometry** found in everyday objects can provide inspiration. You could also set up a still life of objects, or take a trip to a design museum and feel inspired to design from these sources.

A *Some starting points: still life, nature, structures, elements*

Starting points can also be secondary sources such as information found on the internet or books about **designers** (for example, Heston Blumenthal – who works with food; Philip Treacy – a hatmaker who works with a wide range of materials including textiles; and Ron Arad – an inspiring designer of furniture who works with wood, plastics and metals).

Case study

'Cooked to perfection': Heston Blumenthal

The Fat Duck at Bray in Berkshire is a triple Michelin starred restaurant known for unusual dishes, such as snail porridge, basil blancmange beetroot jelly and bacon and egg ice-cream.

Heston Blumenthal, the chef-proprietor, is interested in the science behind cooking, the 'experience' of dining to which he applies his 'molecular gastronomy' technique. He works closely with academics and has built a laboratory at his restaurant, staffed by a food science PhD student. Access to scientific appliances and industrial equipment mean that, for example, liquid nitrogen could be applied to soup, creating soup icicles.

B *Heston Blumenthal*

Case study

Hats, hats, hats: Philip Treacy

Described as surreal and sculptural, Philip Treacy's handmade hats are feats of craftsmanship. He designs *haute couture* and ready-to-wear hat collections at his London studio.

Born in rural Ireland in 1967, Treacy found inspiration from the chickens, geese, pheasant and ducks kept by his mother. While still a student, he made Ascot hats for Harrods. He went on to meet Karl Lagerfeld, then chief designer at Chanel, and designed hats for him.

At the time, hats were not very fashionable, but Treacy decided to 'change that'. His fantastical creations included a replica 18th-century sailing ship with full rigging, and a castle. He often begins by mocking up the shape in straw, then the hat is steamed and moulded on a specially made wooden block.

Treacy also uses more quirky materials for inspiration, including Brillo Pad boxes and photos of faces. He has been designing his own ready-to-wear collection since 1991 and has developed ranges for a number of high street chains, but the heart of his business is still *haute couture* hats.

C Hat by Philip Treacy

Innovation technology: Ron Arad

Ron Arad is one of the most influential designers of our time. The child of two artists, he was born in Tel Aviv in 1951. He moved to London in 1973 to study architecture and made his name in the early 1980s as a self-taught designer-maker of sculptural furniture. He now works across both design and architecture.

Arad defies categorisation and could be described as an architect, a product designer, a furniture designer or even a sculptor.

In 1981, he set up his own company, One Off, with his business partner Caroline Thorman. In 1989, they started Ron Arad Associates in Chalk Farm Road, north London, in the building they occupy today.

D Ron Arad's Voido rocking chair

Case study

Activities

1　a　Set up a still life with objects from home or in your school workshop, then sketch parts of the still life using pencils or, if you are brave enough, use some India ink or paint.

　　b　Use a card window frame to select part of the still life. Base an idea for one of the following products on what you have drawn: fashion bag, piece of jewellery, radio, wall light.

2　Working with your classmates, produce a display showing the work of as many different designers as you can find. Make sure you list their names, the period when they worked, and what they are famous for.

∞ links

For more information on contemporary designers, see **www.designmuseum.org**

For more information about Philip Treacy and his work, see **www.philiptreacy.co.uk**

For more information on Ron Arad, see **www.ronarad.com**

Summary

All designers have different starting points for inspiration. The case studies in this chapter show design working in three different areas: food, textiles and resistant materials.

First-hand reference is a great starting point for designing.

Study tip

These designer case studies will help you to understand a bit about how designers work and what a career as a designer might be like. Make sure you know how a particular designer's work has developed through his or her career, and what the major influences were.

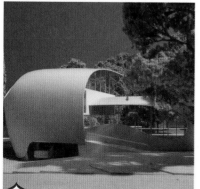
E Arad is also a stylish architectural designer

4.1 The functions of packaging

Packaging

The purposes of packaging can be summarised by I PICT PD (I picked product design):

P **Protection** – packaging protects the product from damage in transit and prevents tampering.

I **Informing** – information for buyers and users, including a picture of the product, can be printed on the packaging. Some information has to be included by law.

C **Containing** – some products are made up of lots of tiny pieces, for example breakfast cereals. Flat-packed products also include many individual components. Care must be taken to keep these pieces together.

T **Transportation** – designers must consider how packaging can best enable products to be packed in bulk for transporting in lorries or containers. How the product can be easily lifted and stored are key issues.

P **Preservation** – many products, especially foods, need to be prevented from deterioration due to temperature changes or bacteria.

D **Display** – attractive and distinctive packaging enables buyers to find the product easily once it is stacked on the shelves. Packaging often advertises the product.

The mobile phone box shown (**A**) might contain the following information:

1 **company logo**
2 name of product
3 information about the phone
4 information about different models of phones
5 European Standards logo
6 barcode
7 website address
8 items included in the box
9 **recycle logo**
10 where packaged
11 picture of the product.

This packaging serves to Protect, Promote, Inform and Transport safely in quantity.

 Outer packaging for a mobile phone

Objectives

Understand that packaging has six functions: protection, information, containing, transportation preservation and display.

Learn how to find additional information about use, safety, appropriate care and storage on the packaging itself, along with information about recycling the packaging.

Key terms

Company logo and trademark: company symbols and service marks used to advertise and display products.

Recycle logo: a symbol used to specify different types of recycling destinations.

Safety logos: identifiable logos on packaging to give the consumer confidence that the product has been tested.

Remember

The mnemonic **IPICTPD** (**I** picked product **d**esign) can remind you of the six functions of packaging: Protection, Information, Containing, Transportation, Preservation and Display.

www.toyysco.sample

Figure **B** shows a child's game that might include the following information on the packaging:

1   company logo and **trademark**
2   name of product
3   information about the game
4   European Standards Logo
5   barcode
6   website address
7   items included in the box
8   recycle logo
9   picture of the product
10   age restriction.

B

Labelling

There is a legal requirement to include certain information on the packaging of some products. This information may relate to **safety** and appropriate use. Some frequently used labelling symbols are shown (**B**). More detail is given elsewhere in this book.

Labelling can also be used to give information about the product. The recycling logo indicates that the empty packaging can be recycled by putting it in the appropriate bins. If the packaging is plastic, the recycling label will identify the type of plastic to help the recycling company.

Legal requirements

It is a legal requirement to carry warnings about the dangers of smoking on cigarette packets. Toy packaging must state the appropriate age of the user. All packaging should carry CE European Standards or BSI British Standards logos, stating that the product has been tested for safety and fitness for purpose.

links

Find out more about food labels at: www.chewonthis.org.uk

Activity

Complete a table of materials used for different types of packaging; for example food packaged using thermoplastics and boards. Then list the uses, advantages and disadvantages of the packaging, and whether the packaging can be recycled.

Study tip

- Learn 'IPICTPD' – it will help you remember the functions of packaging. This will be useful if you have to analyse packaging or write specification points.

- Be able to identify logos and information used on packaging. You might be asked to explain what a logo means or you might be asked to design suitable packaging for a product.

Summary

Packaging has six functions: protection, information, containing, transportation, preservation and display.

Additional information about use, safety, appropriate care and storage can be given on the packaging itself, as can information about recycling the packaging.

links

See 5.3 for more information on labelling; including legal requirements.

Industrial manufacture of packaging

A very wide range of materials from metals through to glass are used to package products. However, different types of paper and board are commonly used.

Paper and card as packaging materials

The choice of paper and board for packaging is wide. There are seven different types of paper and board available, ranging from a thin material to something much thicker and more rigid. See 10.3 for a full description of these.

Printing on packaging

There are several techniques for printing on packaging:

- **Lithography** is the most common form of commercial printing, and works on the principle that water and oil do not mix.
- **Flexography** is a form of relief printing used on packaging labels, tape, bags, boxes and banners.
- **Screen printing** is a printing process that uses a mesh to support an ink-blocking stencil. The attached stencil forms open areas of mesh, and a squeegee is moved across the screen stencil forcing or pumping ink through the woven mesh in the open areas.

A *Examples of lithography*

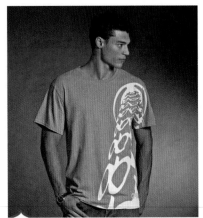

B *Screen printing*

Die cutting and creasing

The outline of a piece of packaging is cut, creased or perforated using the **die cutting** technique. Folding ads stress to board and can make it crack, crease or tear wrongly. To reduce the stress it is scored or creased, creating a crisp clean edge. A die is made from cutting knives and scoring rules.

Printing registration allows the printer to make sure that the surface graphics fully cover the packaging and line up with the die cut box. It ensures that the colour or image printed on the surface overflows the edges or perimeters, allowing a tolerance for cutting and **creasing**.

Key terms

Die cutting: a technique used in the printing process, involving cutting through with a blade attached to a plywood base. This is known as a cutting forme.

Creasing: squashing the card so that it can easily be folded.

Blister packaging: packaging using a pre-formed plastic blister and a printed paperboard card with has a heat-seal coating.

Skin packaging: packaging used to seal a product between a layer of heated plastic and a layer of adhesive coated paper.

C *Flexography*

links

See 4.1 for more information on the functions of packaging (IPICTPD).

See 4.3 for more information on branding.

See 10.3 for more detailed information about the different types of paper and card used in packaging.

■ Types of packaging

Blister packaging

Blister packs are inexpensive, durable, transparent and tamper-proof. Clear plastic blisters are used in conjunction with either a cardboard backboard or inlay to enable the consumer to examine the product. The blister cavity is vacuum-formed to fit the product.

Blister packaging is used on point-of-sale (POS) displays, product packaging and product launches. A good example of blister packaging is medicines in a foil strip.

Skin packaging

Skin packaging involves sealing a product between a layer of heated plastic and a layer of adhesive coated paper. The plastic is trapped in a frame that travels between the oven and the vacuum area. Once the plastic is secured and the frame lifted out of the way, the board is loaded into the vacuum machine and the product placed on top of the paper. The oven heats the plastic for a pre-set time, the frame lowers and the vacuum is activated, pulling the plastic down over the product against the paper. The hot plastic melts the adhesive, and the paper and plastic are bonded together.

Security packaging

- Packaging can be securely sealed by a number of means: security seals, clamshell packaging, induction sealing or holographic stickers. See the Links in the margin for sources of more detail.

Insulation

Polystyrene sheets provide effective insulation for products, as does corrugated card. Bubble wrap and polyethylene foam are also good insulators.

Insulation provides protection for fragile items. It also keeps temperature constant if you are transporting food products or plants for example. Think about how a pizza is wrapped for delivery.

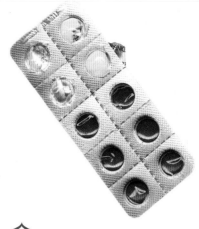

D *An example of a blister-packed product*

∞ links

- To find more about printing go to *www.printingandpackaging.co.uk*
- For more about printing on packaging, see 9.13.
- For more about paper and board-based composite materials, see 10.3.

∞ links

For more about security sealing and insulation, see:
www.packaging-gateway.com
www.aerobord.ie
www.zouchconverters.co.uk
www.phflexible.co.uk

Activities

1. Design blister packaging for a product you have made in school. First ask your teacher to show you some workshop samples.
2. Look at three pieces of packaging and discuss with your peers and your teachers linked recycling issues.

Summary

The wide variety of packaging materials available includes paper, card and plastics.

A number of processes are available for printing packaging materials, including lithography, flexography, screen printing and die cutting.

Types of packaging include blister packaging, skin packaging and security packaging.

Brand development means creating a potentially valuable asset for a company. A 'brand' involves using a name, term, colour, symbol, design, or combination of these elements, to identify the goods or services of a seller, in a way that makes them particularly appealing. Branding can speak powerfully to who we are, what we want and who we want to be. Brands often humanise the product or service and give it a personality of its own.

The power of branding is so strong that we often buy a product because of the adverts on television, the look of the company logo and packaging and the website promotion.

■ Why create a brand?

Branding is a way of making a company's product or service more visible and recognisable. It also gives it a 'personality' with which consumers can engage. By inviting us into this personal relationship it builds up the company's reputation and makes customers feel secure.

Many people stay with a particular brand of car. This could be due to loyalty or to an association of quality with the brand.

The same process works with food products. Think about the food your family buys every week. Does it vary or are particular brands particularly favoured? Discuss with your family why they buy a particular brand.

Objectives

Gain knowledge and understanding of the power of branding and its effect on different consumer groups.

Understand that products are marketed using a variety of techniques, including leaflets, flyers, point of sale, packaging and digital media.

Key terms

Brand development: creating and developing a strong product identity that will appeal to consumers.

Digital media: a form of advertising on the computer.

Point of sale (POS): where the product is displayed for sale. A POS is often part of a new product launch. It may be a unique display stand with key information about the product.

A *Some famous brands*

Creating a brand personality

What values do you want to be associated with your brand? Reliability, safety, tradition? Youthfulness, freshness, friendliness? Inspiration, sophistication, quality? No-nonsense, unfussiness, directness?

A brand personality can be used to:

- identify the business so people know who you are
- create empathy with a target group

- show a level of business professionalism
- bring business to the company by making the company visible.

Brand identity can be used in advertising: banners, brochures, business cards, compliment slips, **digital media**, direct marketing, e-flyers, e-mails, exhibition stands, flyers, letterheads, merchandising (for example T-shirts or mugs), packaging, **point of sale**, signage, staff uniforms and websites.

Activities

1. Make a list of as many brands as you can think of in two minutes. Then choose one of these products and list the different promotional materials used to reinforce the brand.

2. Write down the brand values of easyJet: for example, fun, friendly, forward thinking, upbeat. Now discuss how you feel these key words relate to the power of branding.

B *easyJet branding*

Market research

Manufacturers invest heavily in market research to ensure their products and services are wanted by consumers and therefore will sell. Here is a list for you to check against when thinking about market research for your project.

Your target market

- Who is your service/product aimed at?
- Who are your clients/potential clients?
- What do your potential clients expect from your service?
- How old are they?
- Are they male or female?
- Where do they live?
- Do they care about price and affordability?
- What kind of lifestyle do they have?

Market research will help you find out about your clients' values, expectations and needs, and how your competitors operate.

Activity

3. Find out about how two top brands can join together to promote an innovative idea for an effective campaign across different target markets. Nike has joined with MSN, using joint branding to drive awareness of the Nike football game. Look it up on the link given on this page and find out more.

Remember

A brand identity puts over an organisation's personality, beliefs, values, behaviour, attitude and aspiration.

∞ links

Read about how Nike and MSN are using joint branding and forming an online football game:

http://europe.advertising.msn.com/WWDocs/User/europe/casestudies/nike_29-10-04.pdf

Study tip

To help you remember what branding is, think about a brand you know and like. It's useful to know how a logo helps to create a brand image: so you will need to think about the colour, shapes, images and fonts used to attract consumers.

Summary

A good brand identity sets a company apart from its competitors.

A brand personality invites the consumer to enter into a relationship with the product and producer.

Market research can help companies identify their target market and shape their brand identity.

5.1 Human factors

Of all the factors that product designers need to consider, human factors are often regarded as the most important. Focusing on the abilities and limitations of the end user, human factor issues fall into three groups: **physiological**, **psychological** and **sociological**.

Physiological factors

People's physical capabilities, such as their hand-eye co-ordination, size, strength and stamina, affect the ways in which they relate to products. The study of these factors is known as **ergonomics**, which comes from the Greek words *ergon*, meaning 'work', and *normia,* which relates to organisation. Ergonomics was first used to study how military personnel operated machinery; it has since been broadened to examine the human interface with many different products and environments. A good example of ergonomics can be seen in the design of cars, which drivers must be able to operate comfortably, efficient and safely.

Objective

Understand that social, economic and ethnic groups of people often have specific values and needs, which can be an aid to focused designing: for example disabled, elderly and religious groups.

A *Ensuring that the information is easy to understand is a vital part of the car designer's job*

B *Some information is easier to read digitally whilst analogue displays are easier when the information changes rapidly, such as with a car speedometer*

Psychological factors

Designers also have to remember that our senses constantly provide us with information, such as the height of a step or the heat from an oven door. Many products help the user to understand and react to sensory information of this kind, making the product easier to use.

Colour is important because it can influence moods: it may create a welcoming atmosphere in an entrance space or encourage calm

C *Would you be reluctant to put this in your mouth if it was coloured black?*

D *Disabled access is a top priority when designing public buildings and transport systems*

Key terms

Physiological: relating to the body and its movement.

Psychological: relating to the mind and behaviour.

Sociological: human problems in relation to environmental factors.

Ergonomics: the study of relationships between people, products and their environment.

E *How loud does a personal alarm need to be to hurt the ears of an attacker? How loud does an alarm clock need to be to gently wake you?*

thinking in an office. It can also help the consumer understand the product; for example, safety features may be in red, yellow or orange. White is often used to suggest cleanliness, hence the colour of many kitchen appliances.

Sound can be used to provide audible instructions and warnings. Whistling kettles, reversing trucks and car indicators are all examples of products that provide audible feedback. Sometimes the sound is built into products, such as watches, for reassurance that the product is working.

Sociological factors

The design of larger scale products needs to take account of sociological factors. Adequate personal space affects stress levels and is a major factor in public transport. Access is also a key issue: the ease of climbing steps or opening doors can be difficult for children and the elderly.

⚙ links

For more information on ergonomics, see:
www.ergonomics4schools.com

For more information on a special group of users, see:
www.youngveggie.org

Activity

Create a clearly illustrated factsheet for a product, designed to meet the needs of a special group of users, such as the elderly, under-fives, disabled or vegetarians (see Links), and which would not necessarily be used by the wider population.

Study tip

- 'Human factors' is an important area. Learn the meaning of the words 'ergonomics' and 'anthropometrics' (see 5.2).

- In particular, learn why ergonomics and anthropometrics are important when designing products, and to give examples of how they relate to a specific product.

Summary

Physiological factors include: physical limitations; how the body moves; hand-eye co-ordination; strength, size and stamina.

Psychological factors cover the way our senses respond.

Sociological factors cover special groups (including people with disabilities), personal space, access and safety.

Inclusive and exclusive design

Some products are advertised as **exclusive**, which means that they are designed to exclude certain people. As far as possible, designers should aim to make products **inclusive** and accessible to everyone. Excluding certain groups is sometimes inevitable; however you should try to cater for everyone's needs if you can when designing your own products.

Exclusive design

Everyone is different but, in product design, we often put people into different categories. These stereotypes ensure that products are aimed at specific groups, known as target markets. This helps designers, manufacturers and retailers in the design and sale of a product. Food products often need to take into account people's differences: children, the elderly, athletes, pregnant women and nursing mothers all have different requirements as do vegetarians and people with food allergies.

Inclusive design

The ideal product is one that meets everyone's needs. Whilst it is probably an unrealistic dream, designers should aim to make products accessible to all sections of society and exclude as few people as possible. Because people vary so much in size, many products need to be adjustable to suit a wide range of users, such as office chairs, bicycles and car seats. Ergonomics can help designers produce products that are inclusive.

Objective

Understand that for products to be effective, designers, manufacturers and craftsmen need to take account of a wide range of human factors in an attempt to produce inclusive rather than exclusive designs.

Key terms

Exclusive: excluding people by failing to meet their needs.

Inclusive: meeting everyone's needs.

Anthropometrics: the study of human measurements.

5th to the 95th percentile: the 'normal' range that product designers target.

A *Adjustability is a major issue for some products to ensure they meet the needs of a wide range of people*

B *A range of milk products aimed at specific groups of people*

Anthropometrics

One area of ergonomics is **anthropometrics** – the study of people's size. While research-based information about the size of the human body is useful, it can also cause problems. Imagine how tall people would feel if doorways were designed for the person of average height! To avoid excluding people, designers tend to use mathematical data known as the **5th to the 95th percentile**. Doorways are designed to accommodate the 95th percentile – most people can pass through them easily.

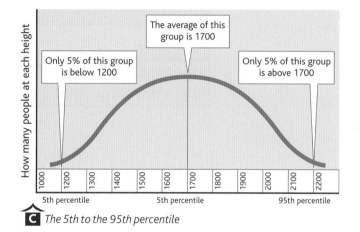

C *The 5th to the 95th percentile*

Ergonomics is also concerned with effective organisation of workspaces, such as kitchens or car interiors. We often use the term 'working triangle' to describe the range of movement needed to perform a certain task.

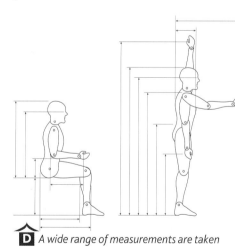

D *A wide range of measurements are taken*

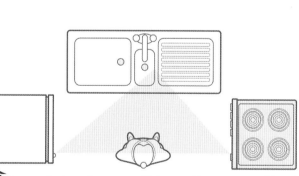

E *Ergonomics is used to create efficient working spaces. Working triangles are a key feature of good kitchen design. Organising the different activities – to reduce the amount of movement a person needs to make – is an important consideration*

Summary

Inclusive design products are designed to be accessible to a wide range of people.

Exclusive design products are designed to meet the needs of a specific target market.

Anthropometrics involves human measurements being used to produce ergonomic products.

⚭links

For more information and ideas on anthropometrics and ergonomics, see:
www.ergonomics4schools.com

Designers and manufacturers should always try to ensure that their products are safe for people to use. When designing your own products you should take into account the wide range of regulations and legislation aimed at protecting the consumer.

Legislation

Many regulations, set up to ensure that products are safe to use, are governed by Acts of Parliament, and you should consider these when designing.

- **The Trades Description Act** – it is illegal to make false claims about a product.
- **The Consumer Protection Act** – aims to prevent the sale of products that may be harmful or defective.
- **The Sales of Goods Act** – goods should be fit for the purpose they are intended.
- **The Consumer Safety Act** – the Government can ban the sale of dangerous products.
- **The Weights and Measures Act** – it is illegal to sell products which are underweight or to sell a short measure.
- **The Food Safety Act** and **The Food Safety Regulations** – guidance is given on food hygiene management.
- **The Food Labelling Regulations** – certain information is required to be shown on most food labels.

British Standards Institution

The British Standards Institution (**BSI**) helps to ensure the safety and quality of products and services. A standard is made up of a series of tests, and companies can pay to have their products tested against national standards. If they meet the standard and the production processes also comply with regulations, they will be awarded the BSI Kitemark. This symbol of quality and safety can be used to assure the consumer that they are buying a consistently reliable product.

European Standards

'CE' marking tells a consumer that the product conforms to minimum European Standards. Most toys sold in reputable shops, for example, carry this symbol, which stands for *Conformité Européene*.

Labelling and packaging

Products sometimes require information to help the consumer use the product correctly. This can be displayed on labels or packaging. Food products provide a good example of this – just take a look in your fridge! Legislation often controls the way in which this information is displayed and helps to ensure that the quality of a product remains constant.

- **Product name** – this must not be misleading. Chocolate flavoured desserts do not need to contain chocolate, but should taste

A *The BSI Kitemark*

B *The 'CE' marking symbol*

links

The British Standards Institute Education website:
www.bsieducation.org

chocolatey! If the label says 'Chocolate Dessert' it must contain chocolate.

- **Ingredients** – this should list all the ingredients in decreasing order of weight, and should also include additives and preservatives.
- **Nutritional information** – this must be provided so that consumers can compare products.
- **'Use by' and 'Best by' dates** – Products that have exceeded their 'Use by' date should be thrown away. Food that is past its 'Best by' date begins to deteriorate but will not be harmful.
- **Storage instructions** – this information must be provided, especially if the product is likely to be stored after opening.
- **Contact address** – the name and address of the manufacturer, distributer or retailer must be supplied so that they may be contacted if necessary.

C The 'e' means that the average quantity must be accurate – the weight of each pack might vary slightly

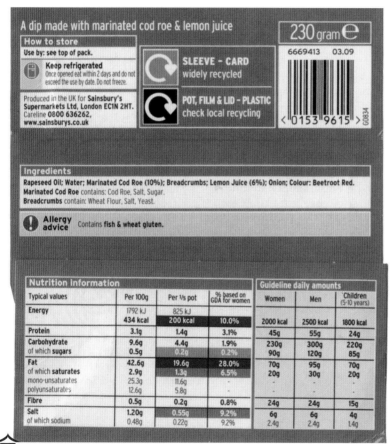

E Food labels should display a variety of information to ensure the safety of the consumer

D The PAO symbol shows how long the product should be kept after opening – for example cosmetics might carry a twelve month recommendation

Study tip

Learn the symbols and meanings for the various quality standard marks.

Activity

Collect labels from a range of products and compare the information given on them. Use these as a resource when producing packaging for your own designs.

Summary

Legislation ensures that products are safe to use.

Labelling and packaging can be used to display information.

The BSI Kitemark indicates that the product meets British Safety Standards.

'CE' marking indicates the product meets European Safety Standards.

Manufacturers must make sure that the products they make are of an acceptable quality to gain a good reputation and to maintain sales. Sub-standard or rejected products can be costly.

Quality assurance

Quality assurance (QA) is an overall approach to ensure that products attain a consistently high standard. Throughout the manufacturing process, materials, equipment, production processes and training of staff need to be checked and monitored. Customer views may also be considered in this process. The BSI Kitemark and 'CE' marking relate to quality assurance as well as product safety. Some manufacturing guilds have their own quality standards and award a quality symbol, such as the Woolmark.

Quality control

As a product or its component parts are manufactured, a series of samples may be taken from the production line and checked to make sure that each part meets a specific, previously set standard. This is known as quality control (QC) and may take place any time during production. Depending on the product, it may be tested after every ten, hundred or thousand have been produced. These tests might include factors such as accuracy of dimensions, weight, flammability, fit and use. Manufacturers should include quality control checks for their own products.

A The Woolmark

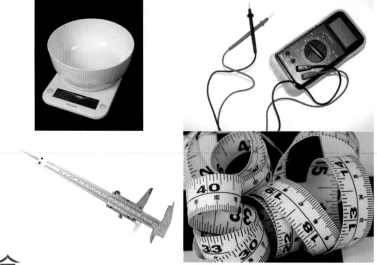

B A variety of equipment can be used to ensure quality

Tolerance

If a large quantity of a product is to be made, it is not always possible to guarantee that every product is absolutely identical. So a **tolerance** may be applied dictating, for example, minimum and maximum measurements. Tolerances are the acceptable range of differences from the agreed standard. These differences are usually documented as plus

UNLESS OTHERWISE SPECIFIED:

WHOLE NUMBER DIMENSIONS:	±1
ONE PLACE DECIMALS:	±0.5
TWO PLACE DECIMALS:	±0.15
ANGLES:	±0.5°

C How a set of tolerances might look

and minus values, and would be applied to factors such as size, weight and performance.

Branding

Brand names identify a company and the products they make, and give consumers an idea of the nature of the product. People often associate particular brands with a certain standard of quality, and it is important for businesses to uphold their brand values (that it, keep the promises they make to their customers).

Consumers looking for quality may show brand loyalty and purchase products based on previous experience or a successful advertising campaign; others may look for further evidence of quality and reliability. The consumers' association's *Which?* magazine tests products against other similar products and grades them against specific criteria. This enables consumers to see which products are best suited to their needs or best value for money, without having to test the products themselves.

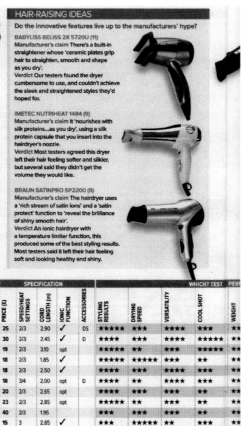

Hot shots

We've had a blast testing manufacturers' claims for their hi-tech hairdryers

If you don't know your ionic breeze from your nutriheat technology, you're in good company. Choosing a hairdryer can leave you feeling like you need a degree in particle physics. So we've made it easier for you to pick a winner, putting 13 hairdryers through their paces in the lab and with a panel of more than 30 testers.

Each of the hairdryers we tested had its good points and bad. Many models have unique features, such as silk protein conditioning or built-in straighteners (see right, or www.which.co.uk/hairdryers for more details).

We also found that some manufacturers' claims are either too unclear to be meaningful, or are overstated.

The top-rated hairdryers are versatile enough for a quick dry or a complex hairdo, depending on your mood. All our Best Buys have a wide range of settings so you can style your hair with the speed, heat and force that give the best results.

You could save a few pounds on our Best Buys by shopping around. We found up to £6 difference by comparing online prices at Argos and Amazon. It's worth doing separate online searches for both the name and the model number because different stockists come up depending on which you use.

HAIR-RAISING IDEAS

Do the innovative features live up to the manufacturers' hype?

BABYLISS BELISS 2X 5720U (11)
Manufacturer's claim There's a built-in straightener whose 'ceramic plates grip hair to straighten, smooth and shape as you dry'.
Verdict Our testers found the dryer cumbersome to use, and couldn't achieve the sleek and straightened styles they'd hoped for.

IMETEC NUTRIHEAT 1484 (9)
Manufacturer's claim It 'nourishes with silk proteins...as you dry', using a silk protein capsule that you insert into the hairdryer's nozzle.
Verdict Most testers agreed this dryer left their hair feeling softer and silkier, but several said they didn't get the volume they would like.

BRAUN SATINPRO SP2200 (8)
Manufacturer's claim The hairdryer uses a 'rich stream of satin ions' and a 'satin protect' function to 'reveal the brilliance of shiny smooth hair'.
Verdict An ionic hairdryer with a temperature limiter function, this produced some of the best styling results. Most testers said it left their hair feeling soft and looking healthy and shiny.

MODEL	PRICE (£)	SPEED/HEAT SETTINGS	CORD LENGTH (m)	IONIC FUNCTION	ACCESSORIES	STYLING RESULTS	DRYING SPEED	VERSATILITY	COOL SHOT	WEIGHT
1 TONI&GUY 2000W Wave Sock Pro Tourmaline TG180D	25	2/3	2.90	✓	DS	★★★★★	★★★	★★★★	★★★	★★
2 NICKY CLARKE Nano Detox Silver NHD011	30	2/3	2.45	✓	D	★★★★	★★★	★★★★	★★★★★	★★
3 PHILIPS TreSemmé SalonControl Ion Shine HP4982/07	19	2/3	3.10	opt		★★★★★	★★	★★★	★★★★★	★★
4 JOHN LEWIS Ionic Hairdryer 5199JU	18	2/3	1.85	✓		★★★★★	★★★★★	★★★	★★	★★
5 REMINGTON Protect & Shine Compact Ionic 2000W D4410	18	2/3	2.50	✓		★★★★	★★★★	★★★★	★★★	★★
6 NICKY CLARKE 2000W Ionic Chrome Hairdryer NCD102	18	3/4	2.00	opt	D	★★★★	★★	★★★★	★★★	★★
7 VIDAL SASSOON Hot tools 2000W Pro Salon Dryer VS536UK	20	2/3	2.65	opt		★★★★	★★★	★★★	★★	★★
8 BRAUN Satinpro SP2200	23	2/3	2.85	opt		★★★★★	★★	★★★	★★	★★
9 IMETEC Nutriheat 1484	40	2/3	1.95			★★★	★★★	★★★	★★	★★
10 BABYLISS Turbo Shine 2000 5529U	15	3	2.85	✓		★★★	★★★★★	★★	★★★	★★
11 BABYLISS Beliss 2x 5720U	35	2/3	1.85	✓	S	★★	★★★	★★★	★★	★★
12 WAHL Power Pik Turbo ZX0S2-801	16	3	3.05		A	★★★	★★	★★	★	★★
13 REVLON CordMagic folding 2000W 93130	19	2	1.85	✓		★★★★	★★★	★	★★★	★★

D *Which? magazine is in many public libraries and is also available online*

∞links

For more information on branding, see 4.3.

See the website of the consumers' association, *Which?*:
www.which.co.uk

The website for the Office of Fair Trading also gives advice on consumer issues:
www.oft.gov.uk

Activity

Study a selection of reports from *Which?* magazine.

Using the same format, conduct your own product analysis of a range of existing products for a product you are studying.

Summary

Quality assurance maintains the quality of manufacturing processes and systems.

Quality control ensures the quality of a product through regular testing.

Tolerances are the acceptable range of differences from the agreed standard.

6.1 Sustainability

Designers, manufacturers, retailers and consumers all have some impact on the environment. Taking account of the materials we use, where they come from, and what we do with them at the end of their useful life is known as sustainability.

Non-renewable resources

Traditionally, designers and manufacturers have used a range of materials from non-renewable resources. This means that the raw materials, such as oil, ores and minerals, are taken from sources that will eventually run out. They are natural resources that have been produced over thousands of years, and so are **finite** resources. The production and disposal of products made from these materials can also have a negative impact on the environment, producing emissions and waste that cannot be broken down.

A *Many products use raw materials from non-renewable resources*

Renewable resources

If we are to make our natural resources last longer, we must aim to use a greater quantity of renewable resources in our products. This will both help to protect our environment and will also help to safeguard the supply of products and services. For example, many timber products now carry the FSC (Forest Stewardship Council) logo, which reassures the customer that the timber supply is from renewable, managed forests. Using such logos allows the consumer to make informed choices.

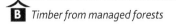

FSC

The mark of responsible forestry

SCS-COC-001259
FSC Supplier

© 1996 Forest Stewardship Council A.C.

B *Timber from managed forests*

Objectives

Be able to take into consideration the environmental and sustainability issues relating to the design and manufacture of products.

Gain a knowledge and understanding of the main factors relating to recycling and/or reusing materials or products.

Key terms

Finite: limited.

Reduce: use fewer raw materials.

Re-use: use a product again.

Recycle: turn the product into a new product.

Biodegradable: break down naturally with the aid of rain and sunlight.

Remember

The six Rs: Reduce, Refuse, Re-use, Repair, Recycle and Rethink can all help sustainability.

∞ links

Check out the website of the Forest Stewardship Council for more information on sustainability:

www.fsc.org

The six Rs

Many people already recycle products and their packaging when they have finished with them. Manufacturers have a moral obligation to develop products that use fewer materials and consume less energy. They should also endeavour to reclaim and re-use the products' component parts when they are no longer needed.

Reduce – **reducing** the amount of raw materials we use will improve sustainability. Designers and manufacturers should aim to use fewer materials and components. Manufacturing processes that reduce the production of toxic substances or emissions should also be considered.

Refuse – refusing to accept extra packaging can make a difference, as can refusing to accept that the easiest way of disposing of products is always the best way.

Re-use – **re-using** products is important because they require very little processing, and therefore have less impact upon the environment than products from raw materials. The reduced production costs can be passed down to the consumer. Milk bottles are a good example, as they only require cleaning and sterilising. Re-using products is becoming more popular and companies such as eBay and Freecycle are helping to encourage the sale of second-hand goods.

Repair – think about whether it is always a good idea to buy a whole new product.

Recycle – many materials can be **recycled**, and recycling symbols on products and packaging inform the consumer how to dispose of them and how they should be processed. Card, paper, aluminium and glass can all be easily recycled. Some plastics can also be recycled although they must first be sorted into the different types using the recycling symbol.

Rethink – perhaps it is time for a radical reappraisal of the way we buy and dispose of products.

How can designers make things better?

Improved sustainability can be addressed by asking the questions in Diagram **D**.

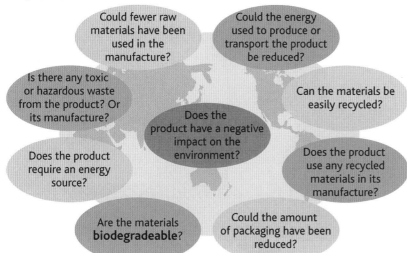

C Most plastics take a long time to break down. So recycling is a better option whenever possible

Activity

Consider a common household product such as a kettle or CD player. What impact does it have on the environment? (Use Diagram **D** to help you.) Is there any way the manufacturer could have improved sustainability?

Summary

Renewable resources can be grown from plants or animals.

Non-renewable resources are used up at a faster rate than they can be replaced.

The 6 Rs stand for Reduce, Refuse, Re-use, Repair, Recycle and Rethink.

D Remember to ask **yourself** these questions when designing your own products

Social impact

It is only recently that the industrialised world has begun to understand the fragile relationship between humankind and the environment. The idea of being responsible for our planet both now and for future generations has become one of the most pressing factors facing designers and manufacturers.

Product lifecycles

The term 'product lifecycle' is sometimes used to describe how long a product will last before it wears out. It may also be used to explain the time taken for a product to become obsolete. Often though, we use this term to show the environmental impact of a product, from the raw materials stage to its disposal (see Diagram **A**).

Key terms

Greenhouse gases: gases, such as those produced by burning fossil fuels, which are linked to global warming.

Fossil fuels: coal, oil and gas.

Activity

Choose a common product, such as a bag of crisps, bar of chocolate or canned drink. Work out, using Map **C**, where the ingredients and materials for the product and packaging have come from and how they might have been transported.

A The product lifecycle

1 Sugar cane and sugar beet are used to sweeten products. Sugar cane is grown in the Caribbean and South America and sugar beet in the UK and Northern Europe.
2 Cocoa beans used to make chocolate are grown in South America.
3 Softwood is commercially grown in North America and Europe. Waste softwood is used to manufacture paper and card.
4 Hardwoods mainly come from trees growing in Europe, New Zealand and Japan.
5 The main sources of Iron ore are Sweden, Ukraine, Australia and North America.
6 Aluminium ore (known as bauxite) is mainly from Australia, Guinea, Brazil and Jamaica.
7 Copper ore is from Chile, USA and Canada.
8 Oil, used for fuel and as the main raw material to make plastics, is produced in the Middle East, South America, Africa and Asia, Russia, UK and Alaska.

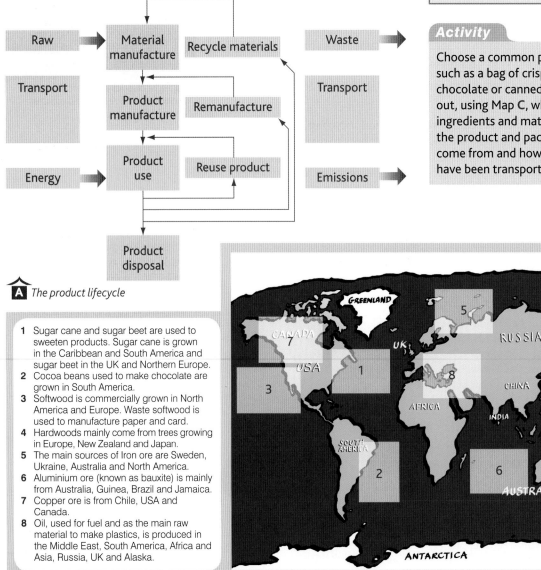

B Where the raw materials come from

Carbon footprint

Everything we produce, buy and use has a carbon footprint: a total amount of carbon dioxide (CO_2) or other **greenhouse gases** emitted during a product's lifetime, including its production, use and disposal. Carbon footprints can have a serious impact on climate change and manufacturers should make every effort to minimise these effects. Reducing the use of **fossil fuels**, cutting down on energy use, careful disposal and the minimisation of transportation can all help to improve the carbon footprint of a product.

Many materials travel long distances before they reach the consumer. Transportation increases carbon footprints and so manufacturers should endeavour to source materials locally where possible, and consumers should be encouraged to buy locally produced goods.

Fair trade

The Fairtrade Foundation, based in the UK, was set up to help alleviate poverty among farmers and workers in areas of the world where previously they had not received a fair price for the goods they produced.

Where fair conditions are promoted and the Fairtrade Foundation's standards are met, products can carry the Fairtrade label. The standards relate to three areas of sustainable development: social development, economic development and environmental development.

Producers and traders must make sure these standards are met by:

- ensuring a fair and stable price is agreed
- investing in projects that enhance social, economic and environmental development
- emphasising the idea of a partnership between traders and producers
- making sure that trading relationships are mutually beneficial and longterm.

The range of products is also growing and some of those now available can be seen in Photo **E**. Composite products, such as clothing, can also be labelled Fairtrade if they are made from a high enough percentage of Fairtrade materials. For example a shirt made from Fairtrade cotton may have non-Fairtrade buttons or stitching thread.

D The Fairtrade Certification mark

E A selection of Fairtrade products

carbon footprint™

C The carbon footprint logo

⦾links

You can work out your own carbon footprint using an appropriate information source, such as **www.carbon-label.com** and the appropriate tool at

the WWF website: **http://footprint.wwf.org.uk** or

the government website: **http://actonco2.direct.gov.uk**

There is more information about obsolescence in 2.1.

Summary

'Carbon footprint' is the term used to describe the amount of CO_2 or other greenhouse gases emitted during the 'life' of a product.

Fairtrade products have been produced under acceptable working conditions and a fair price paid.

The product lifecycle explains the impact of a product on the environment, from its raw material through to its disposal.

Practice questions

1   Products are designed to be easy to use.

(a)   Select one of the hand-held commercial products shown above:

   (i)   Explain the meaning of the term ergonomic.   *(1 mark)*

   (ii)   Explain what ergonomic considerations there might have been when designing the product you have chosen.   *(2 marks)*

(b)   Packaging is needed for the earphones shown along-side.

   This should be a clear blister pack produced from card and plastic.

   Design the packaging so it can be hung on a rack in a supermarket.

   Add notes to your drawings to explain the following details:

   (i)   How the packaging will be hung.   *(2 marks)*

   (ii)   Where the name and company logo will be printed.   *(1 mark)*

   (iii)   Where the instructions for use will be printed.   *(1 mark)*

   (iv)   How the separate parts of the packaging will be assembled.   *(2 marks)*

   Marks will be awarded for quality of communication.   *(3 marks)*

   (v)   In commercial production what process could be used to cut out the card part of the blister pack?   *(2 marks)*

   (vi)   Name a suitable commercial printing process for the card part of the blister pack.   *(2 marks)*

2   Designers often use nature to help them create new products. An example of a product inspired by nature is shown.

The biceps and triceps muscles hold the arm in place, similar to the tension springs on the anglepoise lamp.

Examples of natural forms are shown in the image board alongside:

Select a product from the following list:

Decorative vase	Child's skirt	Child's toy	Food packaging	Clock

(a) Give a target user who might buy your product.

List 3 design criteria which your product will need to meet to be suitable for the target user. *(6 marks)*

(b) Draw one detailed idea which should be made up from natural forms.

Marks will be awarded for:

(i) creative use of natural forms from the image board

(ii) using colour, tone and texture to make your design look realistic

(iii) annotating your design to explain how the main features link to natural form. *(18 marks)*

(c) Explain why your design is suitable for that target user. *(6 marks)*

3 (a) Select three of the symbols below.

Explain what each means and give an example of a product which might display it. *(3 x 2 marks)*

(b) Explain what the British Standards Institute (BSI) does. *(3 marks)*

(c) Explain the advantages of using biodegradable materials for packaging. *(3 marks)*

(d) Explain why recycling is an important consideration for consumers. *(3 marks)*

(e) Using examples of specific products, explain the term 'carbon footprint' and its importance in product design. *(6 marks)*

Processes and manufacture

Knowing how products are commercially manufactured is an important part of the GCSE course in Product Design. In this chapter you will start to understand that the products which we use everyday have to be produced in different quantities, and using different techniques. One of the first issues product designers need to consider is the scale of production. This means the quantity of products which will need to be produced. For example, the method of production would be very different between making a batch of six chairs to making 200,000 identical chairs. We will look at this in more detail later and consider different processes and how manufacturing is organised in different ways.

In this section, you will also start to gain knowledge of some of the key manufacturing terms which you are expected to know about. Manufacturing processes can be grouped into six different areas and you will need to understand how each of these can be applied to the materials you are working with:

- moulding and casting
- forming
- wastage
- conditioning
- assembly
- finishing.

These are dealt with in some detail in the following pages and it is important that you can describe actual examples for each of these terms.

It is also hoped that you will gain a wider understanding of how manufacturing works and how computers are vital to the processes as many modern manufacturing plants are fully automated with few people actually overseeing and maintaining the equipment. Even though many schools have some very sophisticated manufacturing facilities such as laser cutters, milling machines, routers and lathes these are really only used to prototype your ideas or to make very small quantities. The written examination paper will test your understanding of school-based CAD/CAM but industrially computers are often integrated into every function, from placing the initial order to organising the delivery of the products to the customers. Computers also allow the customer to make considerable choices about specific design features – when ordering a new car, for example. Offering a great deal of flexibility to the customer but keeping all of the cost benefits of mass production has only been possible because of the advances in information and communication technology.

This section gives us the opportunity to take a really good look at some of the everyday products we use and broaden our knowledge of how these may have been manufactured. It will also allow us to reflect on the considerations which the product designer would have had to take into account throughout the design and development stages.

Activity

Try to identify the scales of production and the methods of production which might have been used in the products shown below. Then look through the rest of this section of the book and see how much you already know.

7.1 How many products?

Products are made in a range of quantities and the numbers of products made at one time is referred to as the **scale of production**. Common production scales are one-off production, batch production, mass production and continuous production.

One-off production

One-off production involves designing and making a single product, such as a wedding cake, a designer suit or an individual chair. A one-off product is designed and made for a specific purpose. For example, a customer may want something special or a client may ask for a product to be designed within certain performance characteristics.

A

B

C

When a single product is designed and made, all the costs, designing, making and transportation, are built into the cost of that one product – which usually makes the item expensive. One-off production is slow, which adds further to the cost.

Typically, one-off production (for example the ring shown in Figure **A**) involves highly skilled workers who undertake a broad range of tasks using generalised tools and equipment.

Batch production

Batch-produced products are identical and made at the same time in either large or small numbers. Once these products have been made, more of the same products may be made using the same equipment. Chairs, magazines, books and small electrical products are usually batch-produced.

Typically, **batch production** (Figures **B** and **C**) involves some division of labour (different people doing different tasks). Workers might not

∞links

The following pages (7.2) tell you about another production scale – just-in-time production.

have such a broad range of skills. Manufacturing aids (for example, jigs and formers) are commonly used so that each part can be repeated easily. Specialised tools and equipment are even more commonly used.

Mass production

Mass production involves the product going through many stages of a production line. The workers and machines at each stage are responsible for making certain parts of the product. Workers tend to specialise in a small range of tasks.

The cost of the plant (factory or manufacturing complex) is high to ensure that products are made very efficiently.

Mass production was first used by Henry Ford in 1908, for the manufacture of the Model T Ford. Cars are still produced in a similar way.

Key terms

Scales of production: refers to the number of products made at any one time.

One-off production: the making of a single, unique product.

Batch production: when a larger number of products are produced at the same time.

Mass production: manufacturing in high volume.

Continuous production: highly automated manufacture that runs continuously.

Continuous production

In **continuous production**, the products are produced over a period of hours, days, weeks or even years. The production line never stops, running 24 hours a day, 7 days a week. Continuous production usually relies on high levels of automation. Very few workers are used in comparison with the number of products made. Typically, the workforce is less skilled due to the high levels of automation. The cost of the plant is high and is often designed to produce a very limited range of products.

D

Activity

Look at the images on these two pages and explain how each is manufactured in industry. Then look at a product that you are designing and discuss with your teacher which method of manufacture would suit it best.

Summary

Production can be organised in varying quantities, called scales of production.

One-off production involves making a single, individually designed product.

Mass production involves organising a production line so that large numbers of items are produced and each stage of production is organised in an efficient manner.

Continuous production involves high set-up costs, but economies of scale due to the enormous volumes produced.

Study tip

Make sure you understand how scales of manufacture differ. Ensure that you can name at least one specific product that would be suitable for each scale of manufacture.

kerboodle

Organising production

Just-in-time (JIT) production

What is it?

Just-in-time production (JIT) has been developed for products where there are lots of different options available. Modern car manufacture is usually organised round JIT. Each part of the product is planned to arrive on the production line just in time for its assembly. So, if you order a silver car with a red interior, these parts are planned to come together at the correct time.

How does it work?

JIT involves working closely with suppliers. The company upholstering the car seats needs to know exactly how many are needed and when. It is the company's job to organise their delivery to the production line. This is often known as 'logistics'. JIT relies on ICT. As soon as a customer orders a particular product, this information is shared among the suppliers.

Advantages and disadvantages of JIT

- The main advantage is that less money is tied up in the storage of parts.
- The disadvantages include the late deliveries of goods, which stops the production line.

A *A wide variety of products can be manufactured using just-in-time production techniques.*

▓ Electronic data interchange (EDI)

What is it?

Electronic data interchange (EDI) involves the transfer of structured data from one computer system to another without human intervention. For example, information and communications technology (ICT) can be used to transport data files between trading partners – such as a retailer and a manufacturer.

Why is it used?

One advantage of EDI is that it can shorten the supply chain by speeding up the time given to initiate and agree customer orders.

Imagine working in a company that assembles cars. Details of which parts are required when, and how many, are sent electronically using EDI, to the factories that supply the necessary components. The supplying factory will then pass the information directly into its own computer system, which will be updated automatically.

Case study

Textiles: an EDI

An international chain of clothing superstores has special tills with an electronic point of sale (EPOS) facility. These tills gather information about how many garments have been sold throughout the chain and calculate what stock each store will need the next day. Delivery instructions go direct to suppliers, and lorries begin delivery. EDI allows information to be sent quickly to everyone involved including designers, manufacturers and marketing staff.

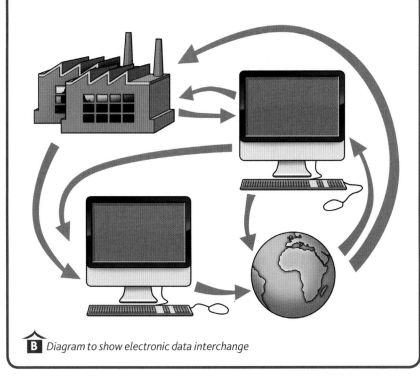

B *Diagram to show electronic data interchange*

What is stock control?

Managing stock

Managing stock effectively is important for any business. **Stock control** requires careful planning to ensure that the business has sufficient stock of the right quality available at the right time.

Stock can mean different things depending on the industry in which the company operates. It includes:

- raw materials and components from suppliers
- work in progress or part finished goods made within the business
- finished goods ready to dispatch to customers
- consumables and materials used by service businesses.

Imagine working in a shoe shop. When the customer goes to the till to pay for their shoes, a point-of-sale till will monitor which styles and sizes of shoe have been sold by scanning the barcode. The computer system can monitor when shoes might go out of stock and respond rapidly by sending EDI to request more.

In order to meet customer orders, products have to be available – although some factories are able to arrange deliveries just in time. If a company does not have the necessary stock to meet orders, this can lead to a loss of sales and a damaged reputation. This is sometimes called a 'stock-out', and may be experienced in various industries; for example, the food industry (affecting restaurants or supermarkets) and the textile industry (affecting high street clothes stores).

Objectives

Understand how ICT can be used to control and monitor stock levels.

Understand how ICT can be used to share information about products all over the world.

◯◯ links

For more information about EDI (electronic data interchange), see 7.2.

Key terms

Stock control: managing the amount of stock held, by monitoring ordering and outflow.

Product data management (PDM): a form of computer software used globally in business process automation.

A Stock control

Activity

1 Draw a diagram explaining how stock control would operate for the product that you are planning to design and manufacture.

It is important that a business either holds enough stock to meet anticipated orders or can get it in time. For a high street retailer, this means having sufficient products on the shelves to meet demand. However, holding too much stock is expensive. Disadvantages and risks in holding stock include:

- storage costs, such as the rent payable for a warehouse
- bank interest, if the stock is financed by a loan
- risk of damage by fire, flood or theft (most businesses insure against this and pay regular insurance premiums)
- stock may become obsolete if buyer tastes change
- some stock (particularly food) may perish or deteriorate.

B *We can all dream of the perfect world, with the best possible systems working perfectly*

Product data management (PDM)

The right tool for the job

Product data management (PDM) is a term used to describe the control of data in business process automation, allowing businesses to manage complex product development that can be shared across many organisations and globally. PDM is like an electronic filing cabinet that each different development process can access. For example, if the price of fabric goes up, this information will be sent automatically to the costing department; if the cost of plastics is reduced or the weight of flour has changed then all this information is fed back to the correct development team globally or locally. The advantage of PDM is that products can be manufactured globally and monitored by the design team.

Advantages of PDM

- Many different aspects and key business processes (such as product enhancement requests and engineering change management) may easily be managed, automated and revised from one central location.
- Secure collaboration between internal employees and global suppliers throughout the entire product lifecycle may be used to increase productivity.

Summary

Managing stock levels is important for the success of any business.

Computer technology, including barcode scanners, EDI and product data management (PDM) can all contribute an enormous amount to stock management.

8.1 Computer aided design and manufacture

■ CAD and CAM

Computer-aided design (**CAD**) can create, modify and communicate product ideas. Computer-aided manufacture (**CAM**) is used in various manufacturing processes to monitor and control production. CAD/CAM is a system incorporating both techniques.

Computer systems have three elements: input (CAD), process, and output (CAM). At school you use CAD on a desktop publishing system for input; the information is then sent through a computer numerical control (**CNC**) process. You use CAM hardware – such as a printer, laser cutter, miller or router – to output your product.

CAD in practice

Some advantages of CAD/CAM are:

- accuracy
- the potential for storage and use of ideas and information
- repetition
- less human intervention and reduced labour costs
- the capability for full automation
- flexibility and the facility for quick change set-ups
- saving of planning and development time, allowing more time for production.

You can use CAD in school to:

- make templates to draw around on materials for cutting
- improve the accuracy of your drawings
- create the numerical data needed for the use of CNC machinery.

A A design with CAD input

B A student uses a laser cutter to cut sheets of acrylic as part of a CAD/CAM process

CAD can also be used to:

- calculate the nutritional profile of a food product
- work out manufacturing costs and suggest a retail price
- scale up recipes for commercial manufacture
- take digital photographs of proposals.

Alternatively, CAD can be used for mathematical and engineering calculations such as **stress analysis**. This is very important for buildings design, and for the manufacture of products such as cars.

There are also disadvantages to this way of working. They include:

- difficulties with ensuring data is secure
- risk of data corruption
- high initial investment in plant and training.

■ Rapid prototyping

This fast, accurate process enables products to be modelled from the designer's CAD drawings, which are passed directly into CNC machinery. These models provide information for both designer and manufacturer; problems are identified early, thus avoiding unnecessary delays and costs.

Rapid prototyping is used to replicate injection-moulded components, for example where the cost of making the mould might run to hundreds of pounds. It allows complex forms to be tested for fit prior to investing in costly mould-making.

Stereolithography

Stereolithography (STL) is a rapid prototyping process that produces realistic models and working prototypes. A CAD/CAM laser draws outlines of the product onto liquid resin. Where the laser touches the resin, it solidifies, building a 3D prototype identical to the drawing.

Rapid prototyping at school

In schools, rapid prototyping allows complex forms to be made without the need for 'making' skills. The two main systems used in schools are a powder system which involves 3D printing. Each layer is printed onto a bed of powder (for example a Z-Corp machine). The other system extrudes ABS images. The first method is cheaper but the prototypes are not as durable. The ABS version can be achieved using a 3D prototyping machine.

Rapid prototyping in schools requires a high level of skill in CAD, and the prototypes cannot be altered easily without being completely remade.

Summary

Computers have three systems: input (CAD), process and output (CAM).

CAM can be used to monitor and control production processes.

CAD can be used for rapid prototyping and modelling.

CAD, CNC and CAM are all used regularly in the school workshop.

C *Rapid prototyping at school*

D *Model produced using RP*

Video conferencing

Video conferencing, using computers, webcams, and communication tools such as Skype, have enabled us to talk 'face-to-face' with others all over the world. This has considerable implications for design and manufacturing: for example, a design team in London can watch their product being made in Thailand.

Remote manufacturing

For many years, design and manufacture were carried out on the same site. Today it is common for designers to send their designs across the world, by electronic means, for manufacture. In book publishing, for example, editorial and design functions may remain in the UK while printing and binding take place in the Far East. The advantage of 'remote manufacturing' is that it is cheaper to print in the Far East and transport books back to the UK for sale. However, the impact on the environment may make this option less attractive in future.

Prototypes and mock-ups

A **prototype** is a highly finished model or functioning product used to demonstrate a design. It is used for testing and may then be considered for manufacture. If you do not have prototyping facilities in your school workshop, you can use card or modelling clay to develop your product.

A **mock-up** is a rough prototype made from materials such as card, MDF or plywood. Designers use modelling clay or polymorph to model products such as handles and to demonstrate ergonomics. Dyson, for example, produced rough mock-ups from card and modelling foam.

Orthographic drawing

Orthographic projection is a means of representing a 3D object in two dimensions. In your coursework it is important to demonstrate your understanding of and skills in orthographic projection. This procedure can then be transmitted to a CNC output machine. Most schools

B A video conference in progress

C Prototype of a car

D An orthographic representation normally includes drawings from several angles

Objectives

Understand how ICT is used in manufacturing beyond basic CAD/CAM.

Understand how CAD/CAM can facilitate global communication.

A Designs produced in the UK may be manufactured far away

Remember

Photograph each stage of the modelling process for your portfolio.

⚭ links

For more information on ergonomics and anthropometrics, see 5.2.

For advice about developing and modelling ideas, see 13.4.

Activity

1 With a friend discuss your file sharing and talk about how you transfer information with others via the internet. Then produce a mind map on a large sheet of paper.

Study tip

Many products are designed in the West but manufactured in the Far East. Why do you think this might be?

How do you think ICT might facilitate this?

have ProDESKTOP software (a 3D modelling packaging) and 2D Design or Coral Draw. These can all produce orthographic projection drawings. Alternatively, you can provide orthographic drawings using paper and a pencil as part of your planning in development.

Modelling in CAD

Realistic 3D models are designed using CAD software. These **CAD models** can be used for the development of an idea. Modelling is an important stage of the design process, enabling designers and clients to consider ergonomic and anthropometric issues.

Software sharing

Millions of people around the world share access to the internet and software. ICT software sharing is crucial in manufacturing: files need to be uploaded and downloaded for manufacture, stock control and the sourcing of data. This enables changes to be implemented immediately anywhere in the world.

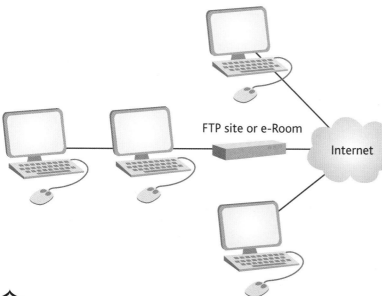

FTP site or e-Room

Internet

E *Software sharing enables data to be distributed globally*

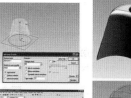

F *Modelling a design in CAD in school*

£19.99

PELLET POUCH

G *The finished product from the design shown in* **F**

Activity

2 Imagine that the product you are designing for your GCSE controlled assessment is to be manufactured globally. Consider all the issues that might need to be discussed through video conferencing. List some questions that you would want to ask your manufacturer.

Summary

Video conferencing and remote manufacturing have enabled companies to spread their workforce globally.

Ideas can be developed through modelling in CAD, prototyping and mock-ups, and orthographic drawings.

Software sharing enables ideas and information to be shared instantly around the world.

8.3 Automating large-scale production

What is automation?

Automation involves the use of computer numerical control (**CNC**). CNC uses computer systems to read instructions, drive and control industrial tools, machines and processes, and reduces the need for human intervention. CNC enables numerous interlinked subsystems to be centrally controlled. It can drive the use of robots or materials handling systems to carry out repetitive or dangerous tasks.

A CNC controller is a powered mechanical device typically used to make components by selective removal of material. CNC technology can also be used to exchange data with other computers and can therefore become part of an automated production system.

Advantages of using CNC machinery include:

- increased productivity, as automated systems can work round the clock
- improved profitability, as different processes and operations can be integrated
- sustained quality, as monitoring and measuring can be continuous.

CNC is very effective when a number of small factories manufacture different components of a product. It can also be invaluable for organising transport, assembly schedules and integration of other operations.

Making products using computers

Robots

Robots can be used for work previously carried out by employees. For example, **CNC robots** may be automated by a central CNC computer to perform tasks on an assembly line. 'Dark factories' use automated robots to run assembly lines for 24 hours a day, allowing the factory's operations to be controlled by computer.

A *An industrial robot on a production line*

Different types of robot are programmed to perform different types of task.

- A first generation robot responds to a pre-set program, doing the same job repeatedly. If this robot were pouring soft drinks into bottles and the liquid missed the opening of one bottle, it would carry on filling the bottles regardless.
- A second generation robot has sensors that relay information to the computer in the factory. These robots can perform a similar

⚭ links

For more information about robots used in manufacturing, see **www.olympustechnologies.co.uk**

You can watch a video of robots on the car production assembly line at **www.metacafe.com/watch/1790456/car_assembly**

automated job on many different products. On a car assembly line, a second generation robot might insert a car windscreen and then the dashboard.

■ A third generation robot has more sophisticated sensors and programming, enabling it to perform many different tasks and giving it 'artificial intelligence'. It not only does the job it has been programmed to do, it can also modify its own program.

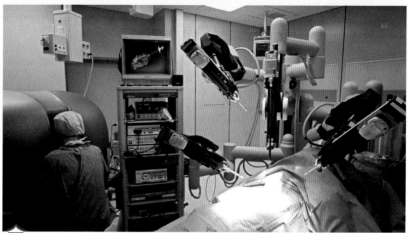

B *A third generation robot performs a surgical operation*

Flexible manufacturing systems (FMS)

Flexible manufacturing refers to the organising of production into 'cells' of machines performing different tasks. Typically, they are laid out in a U-shape, rather than a production line, so that one person can operate several machines. The machines in the cell will be contolled by a central (host) computer that logs the different tasks and operates them in sequence. Often, the system can be used across many different types of manufacturing areas.

An FMS allows savings in time and effort. Most systems like this can react quickly if changes are required to the products being manufactured. An example might include embroidery machines where school logos are printed on polo shirts in different colours. This sort of flexibility makes it easy to cope with small orders.

Activity

With two others, discuss the arrangement of three different products being manufactured using FMS and sketch a system for this operation of injection moulding.

Study tip

■ Robots and automation are used increasingly in commercial manufacture. Think what benefits they may have.

■ Try to write about specific examples of automated manufacturing techniques.

Summary

CNC (computer numerical control) machinery can be used for automated manufacturing.

Robots can be programmed to perform tasks previously performed by employees.

CNC can be used to program flexible manufacturing systems, enabling different products to be made at the same time.

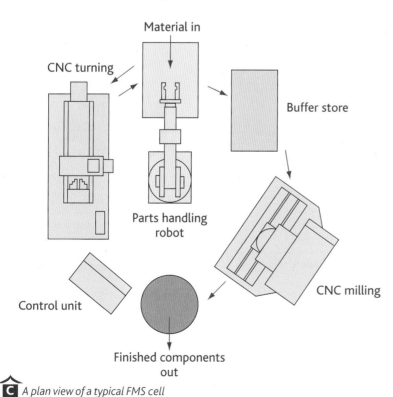

C *A plan view of a typical FMS cell*

9.1 Shaping metals by casting

Casting is a manufacturing process by which liquid materials are poured into a hollow-shaped mould and allowed to solidify.

Sand casting

In this process, a binding agent (sand that feels slightly oily and damp) is used to make the moulds. When the hot metal is poured into the sand cast, the sand holds the metal in place. Example products include manhole covers, metalwork vices, pillar drill tables and engine blocks. The process is as follows.

> **Objective**
>
> Understand that metal components can be made by pouring liquid materials into a mould.

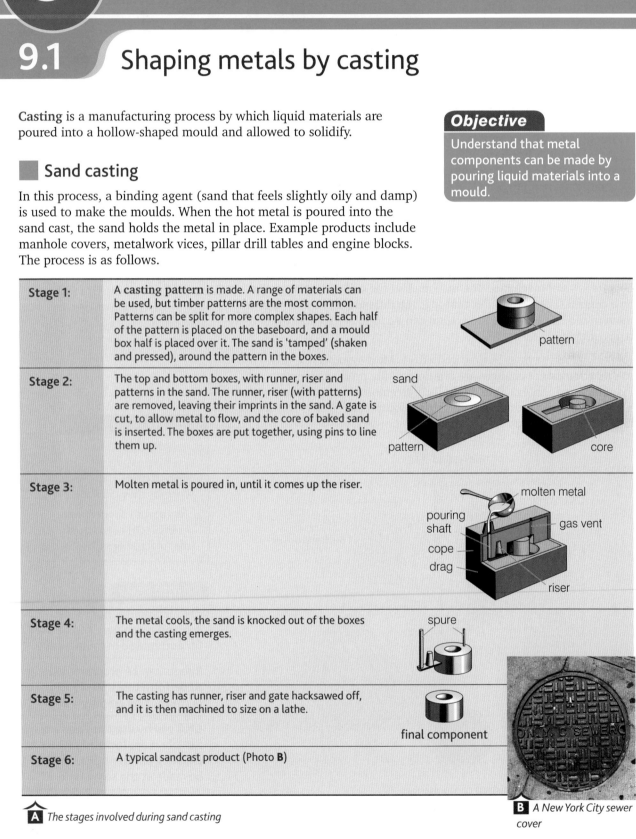

Stage 1:	A **casting pattern** is made. A range of materials can be used, but timber patterns are the most common. Patterns can be split for more complex shapes. Each half of the pattern is placed on the baseboard, and a mould box half is placed over it. The sand is 'tamped' (shaken and pressed), around the pattern in the boxes.
Stage 2:	The top and bottom boxes, with runner, riser and patterns in the sand. The runner, riser (with patterns) are removed, leaving their imprints in the sand. A gate is cut, to allow metal to flow, and the core of baked sand is inserted. The boxes are put together, using pins to line them up.
Stage 3:	Molten metal is poured in, until it comes up the riser.
Stage 4:	The metal cools, the sand is knocked out of the boxes and the casting emerges.
Stage 5:	The casting has runner, riser and gate hacksawed off, and it is then machined to size on a lathe.
Stage 6:	A typical sandcast product (Photo **B**)

A *The stages involved during sand casting*

B *A New York City sewer cover*

Lost pattern casting

This technique is used to form more complex shapes. The pattern is made from polystyrene foam or wax, which is buried into the sand and has risers and runners added. The hot metal is poured into the mould and burns away the polystyrene foam, filling the space with molten metal. Extraction must be used, as the fumes are toxic.

Jewellers use a more sophisticated form of lost pattern casting. The design is first cast in wax. This is repeated many times. Each of these wax models is covered in clay slurry and fired in a kiln. This leaves the hollow shapes of the jewellery ready for casting into. The clay is broken off and the castings cleaned up and polished.

Die casting

Industrial die casting

Die casting is similar to injection moulding and is used to manufacture large quantities of metal products. Alloys with a lower melting point, such as pewter, aluminium alloys and zinc alloys, can be used.

The mould is created by a spark eroding the form required into two blocks of steel. These moulds are water-cooled to control the temperature. The metal is heated in a crucible until molten. A hydraulic ram then pushes a quantity of the molten metal into the mould. Pressure is maintained until the metal has cooled enough for the mould to be opened.

Key terms

Casting: filling a space with liquid material until it becomes solid.

Casting pattern: the shape of the object required, usually made from timber and used to create the hollow shape in the sand.

Study tip

Once you are confident you understand the casting process, concentrate on describing one particular casting technique in detail, using words and diagrams.

Activity

Give three examples of metal casting. Be prepared to explain to the rest of the class which casting process has been used to produce the product.

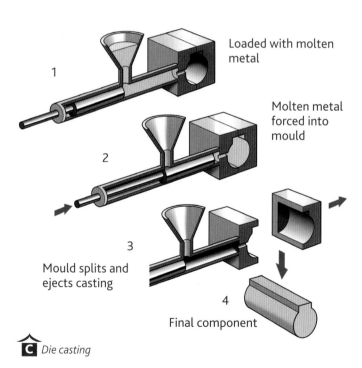

1 — Loaded with molten metal

2 — Molten metal forced into mould

3 — Mould splits and ejects casting

4 — Final component

C Die casting

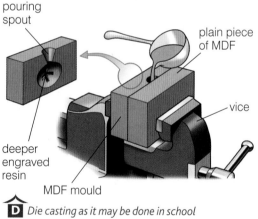

pouring spout

plain piece of MDF

vice

deeper engraved resin

MDF mould

D Die casting as it may be done in school

Summary

Casting is a form of manufacturing that involves filling a space with a liquid material. Sand casting, die casting and lost pattern casting are the most common forms.

Die casting is achieved by forcing molten metal into a mould.

Die casting in school

Die casting can be replicated in school by using pewter melted with an electric paint-stripper gun or blowlamp. The mould can be machined out of MDF or blocks of high-density modelling foam.

Slip casting

Slip casting has been used for centuries to manufacture ceramic products. A plaster of Paris mould is constructed in two or more parts and held together with large rubber bands. The liquid clay, called slip, is poured into the mould and left. As the plaster draws the moisture out of the slip, it forms the wall of the casting, the thickness of which depends on how long the slip is left inside. When a sufficient thickness is achieved, the slip is poured out and the casting is left to dry and harden. The mould is then opened and the casting is removed. Once dry, the plaster mould can be re-used.

There are many uses for slip casting, and examples of products made in this way include teapots, fruit bowls, toilets and washbasins.

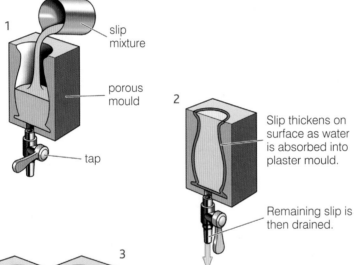

1 — slip mixture, porous mould, tap

2 — Slip thickens on surface as water is absorbed into plaster mould. Remaining slip is then drained.

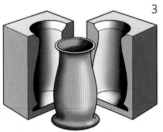

3 — Casting is partially dried and removed from mould. The drain hole is plugged.

4 — After final oven drying, casting is complete.

A The slip casting process

B A wash basin made using the slip casting process

Casting food

Some foods can be cast into shape using temperature and setting agents. Gelatine, a popular setting agent, comes from an animal protein called collagen. Gels can also be made from carbohydrate sources such as seaweed, algae and fruit. The three ingredients used to make a traditional sweet jelly are gelatine, fruit juice and hot water.

Moulding is sometimes used during meringue manufacture to ensure consistent shapes. Factory chocolate begins life as a runny liquid and is moulded into shapes prior to packaging. In the school workshop, chocolate products are made by melting chocolate in a microwave oven or over a pan of hot water (a bain-marie), then pouring it into a mould. Chocolate-covered ice-cream bars also use moulding techniques. Sometimes multiple-moulding stages are required for food products with a more complex design (Twister lollies, for example).

At school you can make your own food moulds using a vacuum former and food-grade polystyrene or PET.

∞links

For more information on finishing food products, see 9.11 (Conditioning food).

Activity

In groups, discuss with your teacher how you could apply casting to a range of materials.

Study tip

You can demonstrate your understanding of casting by giving an example of a material and product manufactured using a casting technique.

C Chocolate before the moulding process

D Casting chocolate into moulds

Summary

Slip casting can be achieved by pouring liquid clay into two halves of a mould.

Casting techniques can also be used for food products.

Different types of moulding

Injection moulding

Injection moulding involves the heating of plastic granules until soft. These are then injected under pressure into metal moulds, where the molten plastic hardens into the desired shape. The mould is then opened and the newly-formed part is removed.

Injection moulding is an extremely versatile and popular form of moulding and the most common process for forming plastics. A large variety of everyday products are injection moulded, including toys, television casings and car dashboards.

The custom designed portion, which produces a specialised product.

This melts and transmits the plastic granules.

clamping unit

mould

heater

material hopper

screw drive

pump piston

This provides the controlled pressure needed to open and close the mould.

1

Plastic pellets fed into heater to melt.

2

Molten plastic pumped into mould.

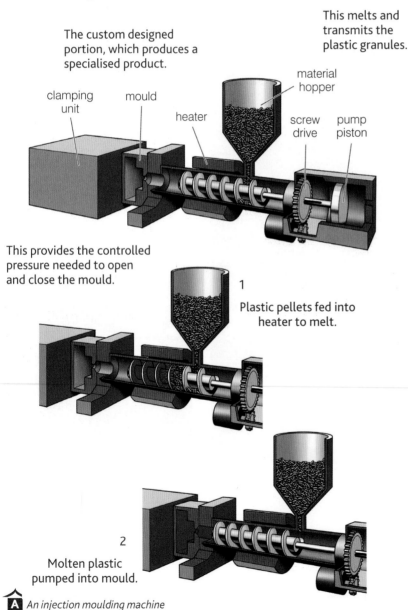

A An injection moulding machine

B Injection moulded sweet container

> **Objective**
>
> Understand that plastic products can be moulded in a single process.

> **Key terms**
>
> **Injection moulding**: a manufacturing process used for the production of plastic objects in large quantities.
>
> **Blow moulding**: a manufacturing process for forming hollow plastic products.
>
> **Parison**: a tube-like piece of plastic with a hole in one end, through which compressed air can pass.
>
> **Rotational moulding**: a process for creating hollow plastic objects.

Blow moulding

Blow moulding is one of the most common moulding processes. It is used to manufacture hollow plastic products such as drinks bottles and shampoo bottles. The blow-moulding machine is based on a standard extruding machine. The molten polymer is fed through a die to emerge as a hollow (usually circular) pipe section called a **parison**. When the parison has reached a sufficient length, a hollow mould is closed around it. The mould mates closely at its bottom edge, thus forming a seal. The parison is cut at the top by a knife, prior to the mould being moved sideways to a second position. In this position, air is blown into the parison to inflate it to the shape of the mould. In the case of carbonated drinks bottles, a parison is injection-moulded in a shape similar to a test tube with a screw thread at the open end. This method of blow moulding ensures that the screw thread is strong enough to contain the pressure from the drink.

After a cooling period, the mould is opened and the final article is ejected. To make production faster, several identical moulds may be blown at the same time. The process is not unlike the one used for producing glass bottles, in that the molten material is forced into a mould under air pressure.

C *PET parisons before blow moulding*

∞ links

For more detailed information on moulding thermoplastics, see:

www.technologystudent.com and click on 'Equipment and processes'.

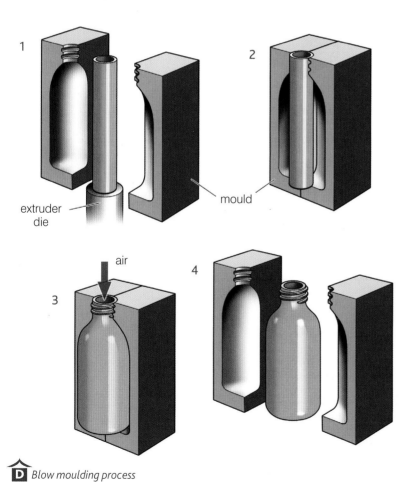

extruder die

mould

air

D *Blow moulding process*

E *A blowmoulded PET/bottle*

Rotational moulding

As an alternative to blow moulding and injection moulding, **rotational moulding** is a process used for producing hollow plastic products. It differs from other processing methods in that the heating, melting, shaping and cooling stages all occur after the polymer is placed in the mould. Footballs, road cones and storage tanks are all made using rotational moulding. The mould is heated and rotated so that the polymer sticks to the inside of the mould. This is cooled and the plastic product removed. Rotational moulding is considerably slower than injection and blow moulding.

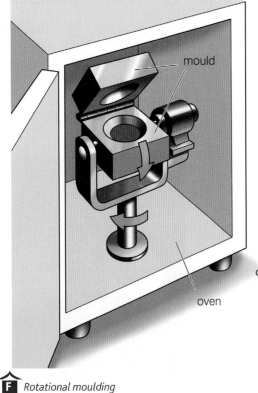

mould

oven

F *Rotational moulding*

Material placed into mould.

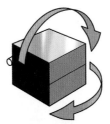

Mould heated to melt material, then spun to create centrifugal force.

Final component

G *Products made by rotational moulding*

Activity

Select and print out images of a wide range of plastic products from the internet. Arrange a display with your classmates to identify which moulding process has taken place for each.

Summary

There are three main methods of moulding thermoplastics.

Injection moulding involves forcing the molten plastic into a mould.

Blow moulding involves forcing the soft plastic parison into the mould.

Rotational moulding involves rotating a heated mould so that the molten plastic sticks to the inside of the mould.

A range of different forming techniques

Vacuum forming

Vacuum forming is used with thermoplastics (for example, for supermarket food packaging). In school, you will probably use high impact polystyrenes (HIPs) to experiment with this process. First, the plastic is heated, then the former is lifted towards the floppy plastic and air extracted using a vacuum pump. Atmospheric air pressure pushes the plastic down onto the former. Finally, the former is lowered, separating it from the product.

Drape forming

Drape forming has many applications. Rolled pastry is pressed into a dish to form the base for a tart. Rolled clay can be formed over a plaster of Paris former. This dries out the clay so that it can be lifted off the former. Acrylic sheets can be heated and formed in a similar way; usually a two-part former is used: the male part of the former pushes the acrylic in one direction, while the female part pushes in the opposite direction.

Drape forming is also used when making hats. Steamed felt is pressed over the former (hat block).

Glass can be formed using drape forming, although in the glass industry it is called slubbing. The glass is heated in a kiln and gravity allows the glass to form over the plaster of Paris former.

Compression moulding

Compression moulding is used with thermosetting plastics. A two-part mould is heated and the plastic is placed into the mould in resin, powder or as a preformed slug. The mould is closed and pushes the plastic into the final form.

Forging

Forging is the shaping of metal using compression forces.

Hand forging

Most hand forging is done with hot metal, usually iron or steel. Traditionally, a blacksmith or farrier uses a hammer and anvil to shape the metal. Most forging ensures that the composition of the metal is undisturbed, making it stronger.

B *Hand forging*

Extrusion

Extruding plastics

Plastics **extrusion** is a process in which raw plastic is melted and formed into a continuous profile. Extrusion produces items such as tubing, window frames and curtain rail. Aluminium can also be extruded in a similar way.

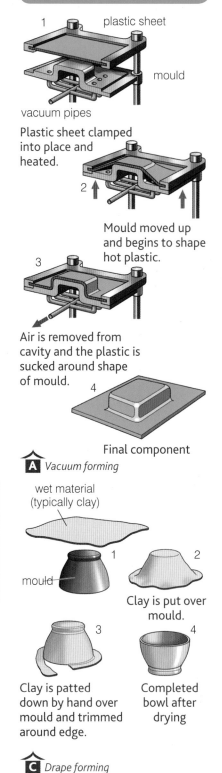

Objectives

Understand that materials can be formed using a wide range of processes.

1 plastic sheet

mould

vacuum pipes

Plastic sheet clamped into place and heated.

2 ↑

Mould moved up and begins to shape hot plastic.

3

Air is removed from cavity and the plastic is sucked around shape of mould.

4

Final component

A *Vacuum forming*

wet material (typically clay)

mould

1

2

Clay is put over mould.

3

4

Clay is patted down by hand over mould and trimmed around edge.

Completed bowl after drying

C *Drape forming*

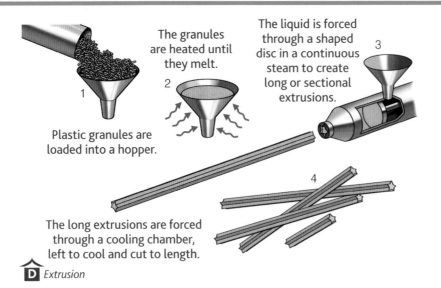

The granules are heated until they melt.

The liquid is forced through a shaped disc in a continuous steam to create long or sectional extrusions.

Plastic granules are loaded into a hopper.

The long extrusions are forced through a cooling chamber, left to cool and cut to length.

D *Extrusion*

Extruding food

When food is extruded it is heated, then pushed through a die. A sudden pressure drop makes the product expand, giving it a light texture. Some savoury snacks and breakfast cereals are created by extrusion.

Line bending

Line bending involves heating a thermoplastic (usually acrylic) sheet over a strip heater until it becomes soft. It is bent to the desired angle – using a jig or former, or by hand. Because the heating is so localised, it makes accurate bending easy.

Sheet metal can be cold bent between folding bars or in a folding machine and this is particularly common when manufacturing white goods such as fridges and washing machines.

E *Using a bench-mounted folding machine*

Pressing

Presses operated by hydraulic rams, which exert massive pressure, are commonly used in manufacturing to bend and form sheet metal. Machine presses are used for shaping all kinds of metals. Products manufactured using a brake press include car body parts.

Summary

Materials can be formed by applying forces to them.

Heating can be used to reduce the amount of force required in the transformation.

Key terms

Vacuum forming: the shaping of thermoplastic by extracting the air from between the former and the plastic.

Drape forming: a technique used for forming sheet materials.

Compression moulding: moulding using heat and a two-part mould to squash the material into form.

Forging: a method of shaping metal using compressive forces.

Extrusion: a technique involving the melting of raw plastic, which is then formed into a continuous profile.

Line bending: the heating and bending of thermoplastic sheet material.

Remember

Presses can be hazardous and must include safety features such as bimanual controls (requiring both hands to operate the machine) or light sensors that stop the machine if the operator is in range.

Activities

1. Think of a product that uses a forming process described on these pages.

2. Working with your classmates, label as many products as you can to explain the forming processes used in its manufacture.

Study tip

- Give an example of a forming process and check you are able to describe its stages and name a suitable material formed in that way.

- Make sure you learn and are able to draw the process diagram for a forming technique. Check that you know what finishing may be needed after forming.

Shearing and die cutting

Objective

Understand the difference between shearing, die cutting and sawing.

Learn about chiselling and planing.

Shearing

Shearing is the cutting or slicing action of a knife blade as it cuts a material such as sheet metal. There are many different types of shears.

- Scissors use the classic shearing action; fabric scissors are often called 'shears'.
- Tinsnips are used to cut small pieces of sheet metal.
- Compressed air shears are used in factories.
- Bench-mounted shears provide more leverage for metals.
- Food processors provide a high-speed shearing action.
- Knives produce a shearing force when used with a cutting board or mat.

Die cutting

Die cutting works on a 'press-knife' principle – like a pastry cutter, for example. Die cutting is used in many industries to cut sheet materials, such as card, fabrics, soft plastics and leather. In the packaging industry it is known as a 'cutting forme', and the blade is usually set into a plywood sheet.

As well as simple cutting, the cutting forme also uses a rounded blade to crease the card for folding. Perforated blades are used for tear-off sections and punches used for features such as euroslots. Foam is placed alongside the blade so that the cut card is pressed off the blade.

Simple die cutting tools can be manufactured in school using MDF blocks to hold the blade in place. Making tools like these will increase your understanding of manufacturing in quantity. Talk to your teacher about working to tolerance.

A *Scissors and a food processor: both of these work in a shearing action*

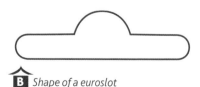

B *Shape of a euroslot*

C *A die cutting forme sliding into a printing press*

Activity

Make a simple die cutting tool to produce a swing label. Keep this safe and you could use it again as part of your controlled assessment project.

Sawing

The principles of **sawing** are the same, regardless of what is being cut. The teeth of a saw are shaped so that they remove a small amount of material on the forward stroke, and some blades are held in tension within a frame to make them easier to control (for example, hacksaws).

■ There are many different types of saw for different materials and tasks, as shown in Photos **D–K**.

D Wood cutting saws often have the handle fixed directly to the blade

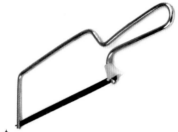

E Hand powered saws cut on the forward (push) stroke

F Powered saws work on several different movements

G Circular saws, used for cutting timber and plastics, rotate the saw blade and the material is moved across the blade

H Powered hacksaws use a forward/backwards motion that copies the manual version and is driven by a crank slider mechanism

I Bandsaws rotate a continuous strip of saw blade

J Jigsaws are used mainly for cutting sheet timber (and occasionally plastics and metals). They move the blade up and down in a reciprocating motion

K Scroll saws also use a reciprocating motion for cutting sheet timber, plastics or metals. The blade is held in tension and moves up and down through a table, which can be angled

∞links

For more information about woodworking tools, see:

http://woodworking.about.com

Key terms

Shearing: a cutting and slicing action.

Die cutting: a method of cutting and creasing material using a simple press knife principle.

Sawing: a method of cutting materials with a toothed blade.

Chiselling: a process used for chipping away pieces of timber, metal or concrete.

Planing: shaving thin layers of timber from the surface.

Chiselling and planing

Chiselling is a manufacturing process for timber and metals. Wood chisels are used for timber and cold chisels for metal. Chiselling involves a wedge-shaped cutting action.

Chiselling wood

There are four types of wood chisel. A sharp edge is essential for them to work properly by slicing across the grain. Chiselling actions include horizontal and vertical paring and chopping; mortising machines can be used to cut deep recesses for joints.

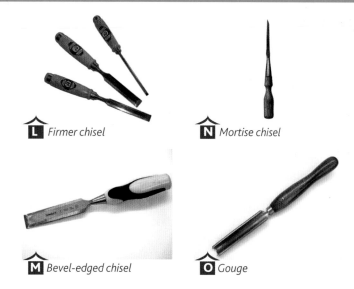

L *Firmer chisel*

N *Mortise chisel*

M *Bevel-edged chisel*

O *Gouge*

Chiselling metals

Cold chisels are made from steel that has been hardened and tempered at the cutting edge. The other end of the chisel is soft to withstand hammer blows.

Planing

Planing has a wedge-shaped cutting action; it is used to shave thin layers of timber and can also be used on some plastics. Many schools use power planes to plane timber, which is a common process in timber industries, using rotary cutters. Hand-held planes are also available.

P *A power plane*

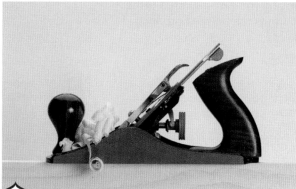

Q *A hand-held plane*

Summary

Press knife tools are used to cut a wide variety of sheet materials.

Chiselling is used to cut out difficult shapes, particularly rectangular hollows.

Planing is a process of shaving off small layers of material.

Drilling

Drilling is the process of making cylindrical holes in solid materials by rotating a drill or boring bit. The drill bit (cutter) must be correctly matched to the material. Most drill bits are made from carbon steel or high-speed steel (HSS). Tungsten-tipped bits are used for drilling brick, ceramics and glass.

The drill bit is rotated clockwise and pressed onto the surface to cut and remove waste material. Drill bits vary enormously in design, depending on the materials being cut.

Milling and routing

In milling, a powered tool with a multi-toothed cutter is moved over the surface of a material to shape it. **Routing** uses a similar technique on woods.

Milling

The material is clamped onto the machine bed, and the cutter raised or lowered (on the 'z axis'). The machine bed can be moved left and right (the 'x axis') or front to back (the 'y axis').

Traditional milling machines move the axes manually, but movements can be controlled using Computer Numerical Control (CNC). The product is designed on screen and the data exported using CNC. The machine outputs the finished product speedily and accurately.

Objectives

Understand the difference between milling, routing, laser cutting and turning.

Understand the processes involved in drilling and abrading.

Understand that CNC can make these processes more efficient.

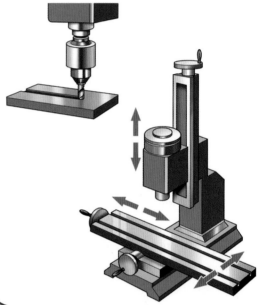

A Milling machines can move the material for cutting in three different directions

Activity

Have a good look around your school workshop – is there a milling machine? Discuss with your teachers how the milling machine works and, if possible, have a go at operating the machine.

Routing

Shapes can be cut using a powered router, following a template. **CNC routing** uses a computer-controlled router in a similar way to CNC milling.

Turning

The material is rotated against a blade or lathe. Wood, metals and some plastics can be turned.

Lathes

The work is held in a 3-jaw chuck, and rotated towards the cutter. The rotation speed is controlled by an adjustable gearbox. The toolpost holds the cutting tool and can be manually controlled or set at a feed rate. The cut can be along the length of the piece, or as a groove into it.

B Wood being turned on a lathe

Woodturning lathes

Woodturning lathes differ from centre lathes: the cutting tool rests on a support and is guided by hand. The wood can be rotated between centres or screwed onto a faceplate.

CNC turning on a lathe

CNC ensures speed and accuracy. The speed of the chuck and the feed of the tool are controlled by a computer program. Costs are reduced, design elements minimised and objects can be produced in large numbers.

Abrading

Abrading involves cutting away tiny particles. Emery cloth is used on metals or for finishing hard plastics. Silicon carbide paper is finer, and can be used dry or with water. Glass and garnet paper is used mainly for timber, occasionally for hard plastics.

Sanding and grinding machines

Most sanding machines are for timber or hard plastics. Linishers are sanders for metals. All powered sanding machines create large amounts of dust, which must be safely extracted. Fixed sanders can be rotary, or use an abrasive paper disc or revolving abrasive belt powered by an electric motor.

The material being sanded must be held firmly in place, and dust disposed of safely as breathing it in is harmful to health.

Grinding machines use abrasive material formed into discs. These can be very dangerous machines as the discs (called grinding wheels) can shatter if abused. Grinding wheels are often used wet to reduce heat and dust.

Files

Files are used to smooth and shape hard surfaces by pressing and dragging hundreds of small teeth across the materials. There are many different types of files, all doing slightly different **filing** jobs.

C *Using a sanding machine*

Laser cutting

Laser cutters are controlled by computers and use a very intense beam of light to burn its way through the material. In schools, lasers are used to cut plastics, fabrics, paper, board and timber materials. Larger industrial lasers can cut through metals, although water-jet cutting and plasma cutting are alternatives. Lasers are very precise and so very little material is removed.

∞ links

For more information about CNC and machinery, see 8.1.

Key terms

Wastage: a process that removes material.

Drilling: making cylindrical holes in solid materials using a rotary action.

Routing: using profiled cutters to decorate or make consistent shapes in material.

CNC routing: routing controlled by input from a computer.

Filing: a pressing and dragging process to waste away materials.

Laser cutters: tools for cutting, scoring or engraving; they use an infrared beam to laser out waste.

Remember

Safety issues are crucial.

Dust extractors must be fitted to all sanding machines in a workshop for health and safety reasons.

Misusing a grinding machine can cause the wheel to shatter, which can cause damage or injury.

Study tip

Check that you understand manufacturing processes, including any preparation and finishing stages in the process.

Summary

Materials can be formed by various techniques of cutting and shearing to meet the desired shape and finish.

The use of machines allows this to be done with great speed and accuracy, especially if CNC is used.

Assembly: traditional wood joints

Traditional wood-joining techniques

Traditional wood joints can be used to give structural strength to products. They have been developed over many years to provide solutions to the problems of joining timber. Here is a selection of wood joints commonly used in schools.

Butt joint

A butt joint (**A**) is the simplest form of joint – each piece is simply glued together. It is a relatively weak joint with a small area for gluing, but can be strengthened by reinforcing with nails, screws, **dowels** or biscuits.

Halving joint

A halving joint (**B**) is cut by removing half the material thickness from each piece of wood to be joined. It can be used for corner joints, tee joints and cross joints. Variations include removing the material from each piece using a tenon saw and firmer or bevel-edged chisel.

Mortise and tenon joint

The basic mortise and tenon (**C**) is a tee joint. It is easy to remember the part names as tools are named after them (tenon saw, mortise chisel). Try to master the skill of making this strong joint, which is used in furniture. You will need to mark it carefully using a mortise gauge. In commercial production, the mortise is milled out and the tenon is machined with a rounded edge.

Dowel joint

A dowel joint (**D**) is a straightforward butt joint reinforced with dowels and aligned holes for extra strength. It is popular in the commercial production of cheap furniture.

Dovetail joint

A dovetail (**E**) is the strongest joint for box constructions in natural timber. Its strength is due to the shaping of the 'tails' and 'pins', which make it difficult to pull the joint apart; it also looks decorative. A dovetail is difficult to cut by hand and jigs are available for using a special dovetail cutter in a router. This joint is used in drawers and jewellery boxes and in furniture where strength is needed.

Lap joint

This joint (**F**) is a little stronger than a butt joint as there is a bigger surface area for gluing; it is often strengthened with nails. A half lap joint is where two pieces of stock, typically of the same thickness, have half of the material removed so that the two boards fit together perfectly without adding thickness at the joint. These joints work well for right-angle connections and are used in furniture, particularly when putting the backs on cupboards.

> **Objective**
>
> Learn about different types of wood joints and the purposes for which they are most suited.

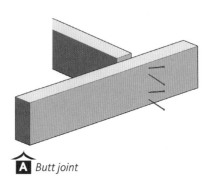

A *Butt joint*

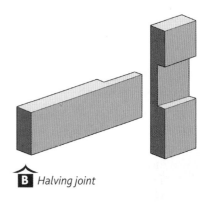

B *Halving joint*

> **Key terms**
>
> **Traditional wood joints**: wood joints that require machining to make interlocking parts.
>
> **Dowels**: circular sectioned pegs made from beech or other hardwoods.

Housing joint

A housing joint (**G**) has a simple slot cut into one piece to increase the glue area. Often made using an electrically powered router, this joint is especially effective with MDF and is used to put shelving together.

Biscuit joint

This gives a quick and easy method of joining boards, either at right angles (as shown) or side-by-side. Slots are made in the board using an electric cutter and 'biscuits' (elliptical pieces of timber that have been dried and compressed) are inserted. When glue is applied the biscuits swell and reinforce the joint in a similar way to a mortise and tenon. This joint (**H**) is used for joining work surfaces together.

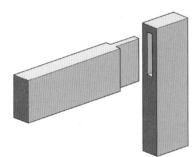

C *Mortise and tenon joint*

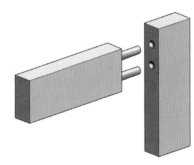

D *Dowel joint*

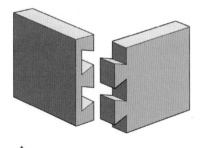

E *Dovetail joint*

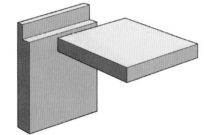

F *Lap joint*

F *Housing joint*

H *Biscuit joint*

Metals can be joined permanently, using heat and a **bonding alloy**. The most common methods used are soldering and welding.

Soldering

Soft soldering

Soft soldering uses a lead-based alloy. **Flux** is applied to the joint and heated with a gas torch or metal soldering bit. This method is used for plumbing joints and lightweight electrical connections.

Hard soldering

There are two forms of hard soldering. Brazing creates a particularly strong joint, so is used for heavier applications. It is often used for joining mild steel, or copper.

Brazing uses a brass-bonding alloy known as spelter, which melts at a much higher temperature than the lead-based alloy used in soft soldering. A borax flux mixed to a paste with water is applied to the joint. It is heated with a gas torch until it turns orange and the spelter melts around the joint.

Silver soldering involves a similar process but using a silver-based alloy. This melts at a lower temperature than brazing spelter, so silver soldering can be used on more delicate materials such as brass, copper and gilding metal.

A Soldering an electronic component into place

Welding

In order to weld two pieces of metal together, the joining edges need to be melted and mixed with the help of a bonding alloy.

Gas welding

This method uses a gas torch to heat up the metals at the joint. A

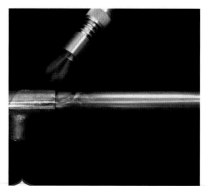

B Soldering copper plumbing pipes with heat from a gas torch

mixture of acetylene gas and oxygen produces a very small, hot flame, which melts both the filler rod and the surrounding metal.

MIG (metal inert gas) welding

This is a form of electric arc welding, often used in schools. It is also used by robots in production welding. An electric spark creates the heat, together with an extremely bright light that can damage eyesight, so it is important to use a protective facemask. The area is cooled with a gas mixture of argon and carbon dioxide. TIG (tungsten inert gas) welding is similar to MIG but is used on stainless steel and aluminium.

Spot and seam welding

Spot welding is a form of **resistance welding** used for joining thin sheet steel. Two metal sheets are sandwiched together, and copper electrodes attached. When a current is passed between them, the resistance creates heat that bonds the metals at a tiny spot. Spot welding is easy to control and ideal for use on a production worked by robots.

Seam welding is a similar process. The material is pressed between two wheels and a continuous weld is produced.

> **Key term**
>
> **Resistance welding**: welding in which the heat is generated by an electric current passing through the join.

∞**links**

For more information about the properties of metals, see 10.5.

For more about manipulating and combining metals, see 11.2.

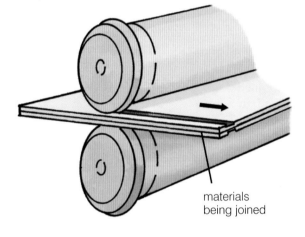

materials
being joined

C *Joining sheets of metal using seam welding*

D *Electric arc welding in progress*

> **Activity**
>
> Talk with your teacher about soldering and welding and think of some uses in school for these processes.

Summary

The most common ways of permanently joining metals are soldering and welding.

Soldering uses a bonding alloy that can be lead based, brass based or silver based.

Welding uses a heat torch or an electric current.

Alternative assembly methods

Joining wood

Traditional wood joints are being replaced by quicker, stronger methods.

Nails, staples and screws

Nails are useful for joining softwood. Hardwoods are more challenging as nails bend under hammer pressure. Glue can be used with nails for strength. In many commercial manufacturing situations nails have been replaced by staples.

Screws are threaded fasteners, comprising a shaft with a spiral groove on the surface and slot at one end for turning. Screws can be combined with glue for strength, and used for metals, plastics and timber. They are commonly found in self-assembly products.

Knock-down fittings

Knock-down fittings are fittings that have been developed for use in flat pack and self-assembly furniture. They include pronged nuts, corner plates and block joints.

Joining metals and plastics

Plastics, including some synthetic fabrics, can be permanently joined using thermal methods such as **ultrasonic welding**. This method, in which high frequency ultrasonic vibrations generate heat, produces a very neat join as no adhesives or connecting components are needed.

Mechanical joining methods are particularly suitable for joining dissimilar materials. **Adhesives** are also widely used.

Nuts and bolts

Bolts and machine screws are made of strong, tensile steel with a continuous thread. They can have a variety of heads and threads, typical lengths being 20–100 mm. Machine screws have the thread over the entire length.

Nuts must have the same thread as the bolt. Some can be tightened by hand while others need a spanner. A washer under the nut keeps it from vibrating loose.

Rivets

Rivets join substances more permanently. Made from metal, they have heads on both sides of the join. Traditionally, the blank end of the rivet is hammered into the second head, though blind rivets (commonly called 'pop rivets') are also seen. A different two-part rivet is used to strengthen pockets on jeans.

Sewing

Hand sewing has been used for joining fabric and leather throughout the ages and around the world. Today, it is even used for some soft

A *Example of a knock-down fitting: a corner plate*

Key terms

Ultrasonic welding: the use of very high frequency vibrations to generate heat within the area to be joined, thereby allowing the materials to fuse together.

Adhesive: a compound that bonds items together.

Activity

1   Have a good look around your home and list all the products that have been manufactured as flat pack. Investigate the use of different screws used in the assembly of the flat pack furniture in your house.

links

For more information about traditional joints for wood, see 9.7.

For information about other ways of joining metals, see 9.8.

Activity

2   Working in a group, test out as many different adhesives as you can and produce a 'Chooser Chart'.

plastics. Early needles were made of bone or wood. Most modern needles are high carbon steel. A sewing machine uses two threads at once: one placed on top of the machine and channelled down towards the needle, while the other is wound onto a bobbin and fed up from underneath. The thread is held under tension to ensure the fabrics are pulled tightly together. Some machines (such as overlockers) use more than two threads.

■ School adhesives

An adhesive is a compound that bonds items together. The table below lists adhesives you may use in school.

Study tip

■ Make sure you can describe how you have joined materials in the school workshop. Remember to describe how you prepared the materials and any jigs or devices you used to help you keep the materials in place.

■ Learn to distinguish between specific systems rather than referring to 'glue'.

B   *School adhesives*

Name	What is it?	How does it work?	Care needed!
PVA (Polyvinyl acetate)	White, water-based adhesive	Soaks into surface, creating a strong bond on setting	
Synthetic resin	Waterproof adhesive	Chemical hardening occurs when mixed with water	Will set in water. Don't wash down the sink as it will block the pipes!
Epoxy resin	Versatile resin adhesive	When mixed with hardener, becomes very hard substance	
Contact adhesive	Solvent adhesive	Both surfaces are coated with this, and allowed to become touch-dry before being placed together	Solvent fumes are dangerous! Ensure good ventilation.
Solvent cement	Clear solvent	Dissolves plastics so they become sticky	Solvent fumes are dangerous! Ensure good ventilation.
Latex adhesive	Rubber-based adhesive solution	Sticks together fabrics and other substances	Smells unpleasant but fumes are not dangerous. Some people are allergic to latex.
Hot melt glue	Usually polythene glue	Heated and placed in an injector 'gun' for easy applying	The glue can be very hot. Take care not to burn yourself!

■ Industrial adhesives

Superglue (cyanoacrylate)

Expensive and difficult to handle, this glue is used to bond hard, non-porous materials. It is sometimes used in surgery instead of stitches.

Polyurethane

Polyurethane (PU glue), used in woodworking and construction, adheres to a range of materials and can expand into spaces between parts.

Summary

Materials can be joined in a number of ways: using fixings (such as screws or nails), adhesives or processes such as welding.

More than one method of joining (for example, nails or screws) can be used in conjunction with another (for example, glue)

The oldest form of joining two pieces of fabric is sewing.

Many forms of adhesive are available in schools. Industrial adhesives may not be as safe to handle.

The term 'conditioning' is applied to a process which changes the properties of a material in order to make it last longer, or to prepare it for a particular purpose. Conditioning can involve heat or cold, chemicals, or other processes such as radiation.

Objective

Understand that materials can have their properties changed through the application of heat, chemicals or mechanical action.

⬭links

For information about food conditioning, see 9.10.

Heat treatment of metals

Heat treatment is the name given to a range of processes of heating and cooling metals, each of them used to obtain different characteristics. The grains that make up the metals change as the temperature increases.

Annealing

Annealing is a process of softening metals to allow them to be bent or hammered.

- Ferrous metals need to be heated to cherry red (about 720°C) for a few minutes and then left to cool slowly.
- Aluminium is rubbed with soap, heated until the soap turns black (at 350–400°C) and then left to cool.
- Copper is heated to a dull red (about 500°C) and then allowed to cool naturally or in water.

A *Heating ferrous metal before shaping it*

Hardening and tempering steels

Hardening a steel involves heating it to 720°C and cooling it rapidly in water. Its hardness is a useful property for tools. However, hard steels can be brittle and may need to be tempered: this slightly reduces the hardness and makes the metal tougher. Many tools – from sewing needles to drill bits – are hardened and then tempered to obtain suitable properties. **Tempering** is done by reheating to a temperature of 230–300°C, then quenching (or cooling) in water.

Kiln firing

Dried clay absorbs moisture and remains crumbly and therefore is unstable and not suitable for most ceramic products likely to be exposed to moisture. Firing clay in a kiln (**kiln firing**) by heating it to around 1 000°C will make the constituent particles fuse together permanently. Coating the fired clay with glaze and firing again makes the ceramic product resistant to moisture.

There are two types of firing:

- Biscuit firing (at 950–1 000°C) fuses the clay.
- Glaze firing is at temperatures of up to 1 300°C, depending upon the type of clay and the recipe of the glaze.

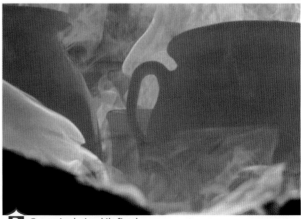

B *Ceramics being kilnfired*

Many metal powders can be fused together in a similar way. This process is known as sintering and is particularly used in bronze bearings. Glass is also fused together in this way.

Conditioning textiles

Conditioning of textiles is to control the moisture content within fibres. However, there are many chemical processes applied to textiles that change the properties of the fabrics.

Many fabrics can be coated to make them waterproof. These fabrics include polyester, taffeta and PVC. Many products, for example tents, can be treated to make the product flame retardant and waterproof. Upholstery can be made to be stain resistant.

Clothing can be conditioned to be breathable, which makes the product comfortable to wear. Most manufacturers of breathable fabrics are based globally, for example in China, Hong Kong, Australia and Taiwan.

Mechanical action such as wire brushing will raise the nap of fabric to give it a softer feel. Stonewashing denim will provide a natural, worn appearance.

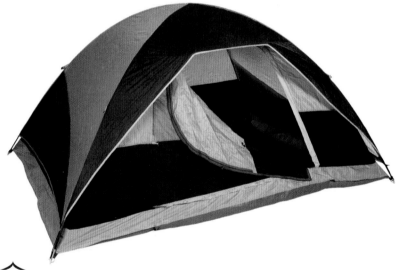

C *Many fabrics can be treated to make them waterproof*

Summary

Conditioning allows the properties of materials to be changed by applying heat, chemicals or mechanical action.

Metals can be heat-treated in a variety of ways.

Clay and similar materials can be heated in a kiln to around 1000°C, which causes the particles to fuse together.

Conditioning textiles can make them waterproof or breathable, stain-resistant or appear worn.

Key terms

Hardening: the heating of steel to 720°C and cooling it rapidly in water to make it harder (also called quenching).

Tempering: a heat treatment technique for metals and alloys.

Kiln firing: a method of 'fixing' clay or ceramics by heating it to around 1000°C.

Activity

Explain the need for the processes of high temperature changes in materials for your chosen area.

Study tip

■ GCSE students should be able to understand how heat is used to condition materials and be able to give examples of products and/or materials treated in this way.

■ This includes understanding the changes to the material properties and why these changes might be needed.

■ Conditioning food using heat

Controlling bacterial growth is a major issue in food production. Temperatures above 121°C destroy all bacteria and spores.

The structure of food also changes during cooking. When baking with yeast, for example, carbon dioxide bubbles from the yeast expand. This causes the strands of gluten to stretch and tangle. The gluten then sets to make the structure of the bread, including a hard crust and holes formed by CO_2.

As well as bonding ingredients together and changing the structure, cooking is used to tenderise ingredients, thicken and preserve, and to change the taste of foods.

■ Food poisoning

According to the Food Standards Agency, **food poisoning** affects around 5.5 million people in the UK each year. It is caused by the consumption of contaminated food or drink, but contamination is difficult to identify because the appearance, taste and smell of the food may not be affected. Although some food poisoning is caused by **toxins** from chemicals or pesticides, most outbreaks result from the toxins produced by bacteria or from the bacteria themselves, which can multiply rapidly given the right conditions (moisture, food, warmth and time). The most common bacteria to cause infection are *E. Coli* and *Salmonella*.

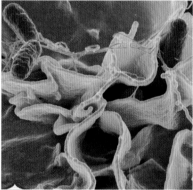

A *Salmonella bacteria invading cultured human cells*

⃝⃝ links

For information on food safety and labelling, see 5.3.

Sterilisation

Sterilisation destroys most bacteria by heating the food to a higher temperature and for a longer period than in pasteurisation. For example, milk is heated to 110°C for 30 minutes. However, caramelisation may affect the flavour of some foods.

Canning

Canning is a form of sterilisation. The food is canned or bottled, then sterilised. It can be sterilised first, then packed and sealed into **aseptic** containers to prevent recontamination. Temperatures required vary – baked beans, for example, are heated to 120°C for 33 minutes before being cooled quickly.

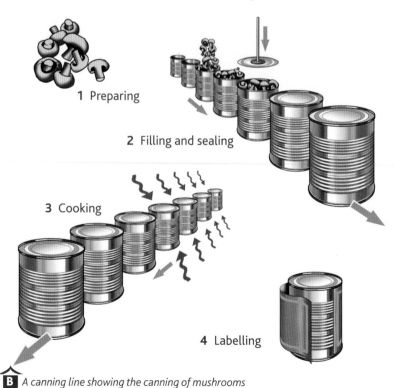

1 Preparing

2 Filling and sealing

3 Cooking

4 Labelling

B *A canning line showing the canning of mushrooms*

Ultra-heat treatment (UHT)

This process involves heating the liquid to a very high temperature for a short time. For example, heating milk to 133 °C for one second kills all bacteria without affecting the flavour.

Irradiation

This method of preservation, introduced in the 1990s, involves bombarding food with ionising radiation (similar to an X-ray). This kills or reduces the effects of micro-organisms, delaying the ripening of fruit and the sprouting of vegetables.

Conditioning food using low temperatures

Chilling

Low temperatures slow or make dormant the growth of bacteria. For short-term storage, food is kept at between 1°C and 8°C. The optimum temperature of 4°C prevents growth of *Listeria monocytogenes*.

Cook–chilling

Ready-prepared dishes, which are cooked and then rapidly cooled to between 0°C and 3°C in 90 minutes or less, will last five days if stored between these temperatures. They must be reheated to 72°C before eating and consumed within two hours.

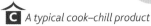 *A typical cook–chill product*

Freezing

Quick-freezing, where food is chilled from 0°C to −18°C in 12 minutes, reduces cell damage. The food must then be stored between −18°C and −29°C.

Accelerated freeze drying (AFD)

Food is quick-frozen, then placed in a vacuum under reduced pressure. When heated, the ice changes to vapour, leaving the food dry. As little heat is used, the flavour, colour and nutritional value of the food are virtually unchanged. Freeze-dried products can be stored at room temperature.

Key terms

Food poisoning: an illness contracted by consuming contaminated food or drink.

Toxins: poisons.

Aseptic: sterile.

Study tip

When developing food products, it is important that you can identify the specific conditioning process needed for commercial manufacture.

Summary

Food can be conditioned in a number of ways to avoid food poisoning. These include heat treatment, chilling or freezing, and irradiation.

High temperatures change the structure of the food and kill off bacteria and spores.

Finishing materials

Finishing surfaces

Many materials need some form of surface treatment to protect them from deterioration and to enhance their appearance. Most plastic products are self-finished by texturing or polishing in order to reduce time and costs.

Objective

Understand the ways in which a range of materials are finished.

Polishing

Polishes such as wax and silicon can be applied to wood, metals and plastics. When **polishing** timber, the polish is applied gradually, in layers, as it slowly fills up the porous surface of the material.

Acrylic may need polishing since its edges are often rough after cutting. The edges can be smoothed using draw filing, then using wet and dry paper by applying pressure and finished with abrasive polish and a cloth or buffing wheel/mop. Abrasive polishes are also used for metals.

Finishing wood

- **Oil** brings out the natural grain of the wood. Teak oil, linseed or vegetable oil can be used.
- **French polish** is made by dissolving shellac in methylated spirit. It is applied using a brush or cloth, re-applied after rubbing down with wire wool, and finished with beeswax.
- **Wood stains** are available in different colours. They provide a surface coating and should be finished using wax or varnish for protection.
- **Sanding sealer** is a solvent-based product used for **sealing** wood. It is applied after wood has been sanded. It seals the wood so that wax or varnish may be applied.

A *Polishing wheel*

Glazing

Glazing agents produce a smooth, protective coating. Materials used to glaze food include:

- **egg** – the white can be used alone, the whole egg or just the yolk mixed with oil or milk to give a variety of finishes
- **milk** – adds shine when brushed over scones and pastry
- **sugar and water** – boiled together for a sticky glaze

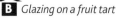

B *Glazing on a fruit tart*

links

For more information on processes such as filing and abrading, see 9.6.

Activities

1. Try to find an example of each of the finishes described. Label each product for future reference.

2. Make a sample board of all the surface finishes available to you in school. Display this in your classroom for further reference.

links

See Spread 11.2 For more information on manipulating and combining metals, see 11.2.

For information about finishing textiles, see 9.9.

- **apricot jam** – warmed and poured over fruit cakes
- **arrowroot** – adds flavour and colour
- **aspic** – a clear jelly, for smoothness and shine.

Fired clay usually requires a glaze to seal and decorate the surface, with firing ranges between 1240°C and 1250°C. Different glazes provide varying colour, texture and opacity.

Paint, varnish and lacquer

Paint, varnish and lacquer are manufactured in oil-based, water-based and solvent-based forms. Paint can be applied to wood and metal, but is unsuitable for plastics. Varnish is clear or translucent and has a matt, gloss or satin finish. They can be applied using a brush or spray can.

◼ Finishing metals

- **Enamel** is a powdered glass mixture applied evenly to surfaces and melted to 1000 °C to form a coating. It can be used for baths, or in a decorative manner for jewellery.
- **Plastic dip coating** involves air being blown through polythene (a thermoplastic powder) to make it behave like a liquid. The metal is pre-heated to 180°C, dipped in and then returned to the oven, where it melts to form a smooth finish. The process is used commercially for products, such as dishwasher racks, and in schools for coat hooks and tool handles.
- **Powder coating** is a more sophisticated, industrial version of dip coating. Powder is sprayed onto the products as they flow through the oven, providing a paint-like finish.
- **Anodising** gives aluminium a durable corrosion-resistant finish, and adds colour. It involves **electrolysis**, using acids and electric currents that are too hazardous for school use.
- **Plating** also uses electrolysis. The thin layer of surface metal gives a durable finish to metals that are prone to corrosion.
- **Hot-dip galvanising** involves coating iron or steel with a thin layer of zinc to provide a resistant finish that protects from the effects of weather.
- **In self-finishing**, the manufacturing process itself gives the material its finish. School chairs, for example, are injection moulded, giving a textured finish on the seat and a highly polished underside.

Key terms

Polishing: a surface application applied to wood, metal and plastic.

Sealing: treating a surface with a solvent-based chemical to protect the surface against damage by moisture or other contaminants.

Electrolysis: using electric currents to transfer particles from one item to the surface of another.

Study tip

Do you understand why some products are finished, and can you select appropriate finishes for specific materials? This will help you, especially if you can also describe the process of finishing a product using a specific technique.

Summary

Many materials need some form of surface treatment for protection and to enhance the appearance of the finished product.

Polishing usually involves some degree of abrasion, as well as applying a protective coating.

Other finishes include sealing, glazing, painting or varnishing.

There are a number of different processes for finishing metals, many involving electrolysis or temperature changes.

When finishing paper and board products, the 'finishing' is done first. This is unlike most other forms of manufacturing.

Objective

Understand the ways in which paper and board products are finished.

Lithoprinting

Offset lithography, by far the most common form of commercial printing, works on the principle that oil and water do not mix. A litho printing plate has non-image areas that absorb water. During printing the plate is kept wet so that the ink, which is inherently greasy, is rejected by the wet areas and adheres to the image areas.

Artwork is produced digitally with graphic design software. An image setter is then used to produce films (either positive or negative). When printing with more than one colour there is a separate film for each ink used. A **photochemical process** uses each film to make a printing plate, the surface of which has non-image areas that absorb moisture and repel ink.

Four-colour process printing

The majority of colour magazines and books, whether printed by litho or gravure, are produced using this process. In the past, the artwork and originals were separated photographically using filters to produce printing plates in four colours: cyan (blue), magenta (red), yellow and black (often referred to as CMYK). Today, this separation is carried out digitally. Because the inks used are translucent, they can be overprinted and combined in different proportions to produce a wide range of colours.

A Lithography is the most common form of commercial printing

Screen printing

The equivalent of the printing plate for the screen printer is the screen – a wooden or aluminium frame with a fine nylon mesh stretched over it. The mesh is covered with a stencil, which is usually produced using a coating of light-sensitive emulsion. The image is printed onto a clear acetate sheet, then this is used as a mask when the mesh is placed under ultraviolet light. This sets the emulsion, and those parts that have not been exposed can be washed off so that ink can be forced through the mesh. The ink is forced through using a rubber strip called a squeegee. In schools, simple masks can be made by cutting paper and placing it under the screen. Screen printing is used on a variety of materials, including textiles, plastics, timber and metal.

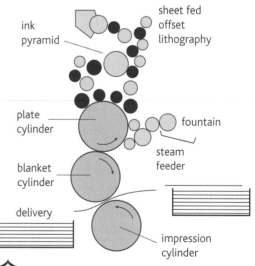

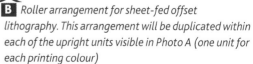

B Roller arrangement for sheet-fed offset lithography. This arrangement will be duplicated within each of the upright units visible in Photo A (one unit for each printing colour)

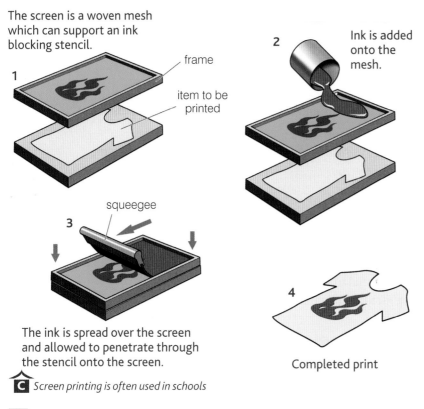

The screen is a woven mesh which can support an ink blocking stencil.

frame

item to be printed

1

2

Ink is added onto the mesh.

squeegee

3

The ink is spread over the screen and allowed to penetrate through the stencil onto the screen.

4

Completed print

C *Screen printing is often used in schools*

Flexography

Frequently used for printing on plastic, foil, acetate film, brown paper, and other packaging materials, flexography (or flexographic printing) uses flexible printing plates made of rubber or plastic. The inked plates, which have a slightly raised image, are rotated on a cylinder that transfers the image to the surface to be printed. Flexography uses fast-drying inks, is a high-speed print process, can print on many types of absorbent and non-absorbent materials and can print continuous patterns (on gift wrap and wallpaper, for example). Applications include paper and plastic bags, milk cartons, disposable cups, confectionery wrappers, envelopes and labels.

Embossing

Embossing is the process of creating a three-dimensional image or design in paper and other ductile (easily shaped) materials. It is achieved using a (female) metal die, usually made of brass, and a (male) counter die: the two fit together and squeeze the fibres of the printing surface. The combination of pressure and heat acts like an iron, while also raising the level of the image higher than the substrate to make it smooth. This is most commonly used on card packaging, but a similar process is used on tissue paper, wallpaper and thin metal containers.

Gravure printing

Gravure involves the engraving of a stainless steel cylinder with thousands of tiny holes in which the ink is collected before being transferred onto paper (or other material). This is an expensive and very high quality printing process. Any alteration to the design requires a new cylinder.

links

For more information about printing processes and finishes, see:

www.technologystudent.com

For more information on offset litho, flexography, screen printing and embossing, search for these terms at **www.about.com**

Summary

Lithography plates are chemically treated to make the image areas resistant to water.

In screen printing, the fine mesh screen acts as a printing plate.

Flexography uses flexible plates made from rubber or plastic.

Embossing involves creating a three-dimensional design by controlled use of pressure and heat.

Practice questions

1   You have been asked to design and manufacture a small decorative gift to sell at
     the summer fair and will need to manufacture 300 identical gifts. Your design will
     need to be simple to manufacture in quantity and can be in any material of your choice.

     (a) Draw your idea for a suitable gift which can be manufactured in quantity using the
          facilities in your school.
          Marks will be awarded for:
          (i)   Quality of notes and sketches   *(4 marks)*
          (ii) Use of tone and colour   *(2 marks)*
          (iii) Feasibility of the idea   *(2 marks)*

     (b) (i)   Name a suitable main material to use to make your design.   *(1 mark)*
          (ii) Explain what finishing methods might be used.   *(2 marks)*

     (c) Explain why your design is suitable for this scale of production.   *(2 marks)*

     (d) Use notes and sketches to explain how you would make 300 of your shapes.
          Marks will be awarded for:
          (i)   An accurate description of each stage of the production process   *(4 marks)*
          (ii) Correct naming of tools and equipment   *(3 marks)*
          (iii) Ensuring all 300 shapes are the same   *(3 marks)*
          (iv) Quality of communication   *(3 marks)*

     (e) Identify a risk involved in the manufacture of your design and explain how you
          would take steps to minimise this risk.
          (i)   Risk   *(1 mark)*
          (ii) Steps taken   *(2 marks)*

2   Manufacturing can be split into distinct areas.

Casting/moulding	Forming	Wastage	Mixing
Fabrication	Assembly	Finishing	

     (a) (i)   Select two of the areas above and give a specific example.
                Selected area 1 …
                Example …   *(1 mark)*
                Selected area 2 …
                Example …   *(1 mark)*
          (ii) Use notes and sketches to explain one of your chosen manufacturing areas. *(4 marks)*

(b) (i) Suggest two benefits of using ICT in commercial manufacturing. *(2 marks)*

(ii) Explain in detail how one application of ICT used in commercial design and manufacture. *(3 marks)*

(c) (i) Select one of the following scales of production and explain what it means. *(2 marks)*

Batch	One-off production	JIT	Mass

(ii) Name a specific product and identify the most appropriate scale of production. *(1 mark)*

(iii) Give reasons why this scale of production is appropriate. *(3 marks)*

From 3544/H 2008

3 Commercial manufacturing requires the organisation of people and production.

(a) (i) Explain the principles of an assembly line system. *(2 marks)*

(ii) Explain the disadvantages to the assembly line workers of this system. *(2 marks)*

(b) Select one of the following products.

Pizza	Board game	Clock	Cushion
Mirror	Mug	Bracelet	Table

(i) Draw your selected product and add notes to explain the separate parts. *(3 marks)*

(ii) Use notes and sketches to show how you would assemble your product. *(4 marks)*

(c) (i) What does the term CNC mean? *(1 mark)*

(ii) Using specific examples, explain how CNC is used in commercial manufacturing. *(6 marks)*

Materials and components

■ Introduction – sources and resources

How did we progress from using stone tools to knowing how to make aircraft and computers? Many of our technological advances have happened as a result of understanding how materials behave, and the development of new materials to solve practical problems. Knowledge of materials has helped create our technological world. Product designers have been able to exploit the properties of materials in new and exciting ways to develop the innovative products that we use every day.

Materials are used to make things. Simple or complex, everything we use is made from materials. A product can be made from one material, or a number of materials can be combined to produce a more complex product. But where do these materials come from? Raw materials are extracted or harvested from the earth and turned into a form that can be easily transported and stored, then processed to produce useable materials.

What is this section all about?

This section focuses on the main groups of materials, and looks at where they come from, the primary processes used to convert them into useable materials, and their stock forms. The properties of materials are explained, including examples of how these properties influence the design of a product. Examples are given in each section, showing how the various materials have been used to produce successful products.

Food groups are explained, with reference to the guidelines on healthy eating, including the importance of following a balanced diet. Information is given about the common forms of food available to the consumer, including methods of preservation. The traffic light labelling scheme, used by some supermarkets and food manufacturers, is also discussed.

The use of electronic and mechanical control components, which are often used as building blocks to produce interactive products, is discussed. Common components are described with examples of how, and where, they are used.

The section also takes a closer look at how materials are manipulated and combined to create new materials, and how 21st century products are being shaped by the development of modern and smart materials.

Throughout the section there are links to further reading, which will help your understanding of materials and their properties.

After you have read this section, you should expand your knowledge of materials and their properties by looking more closely at the products you use every day. Can you identify what materials they are made from? Which properties has the designer targeted to produce a successful product? How was the raw material processed in order to manufacture the product? Do you think the materials used in the product came from a sustainable source, or are they suitable for recycling?

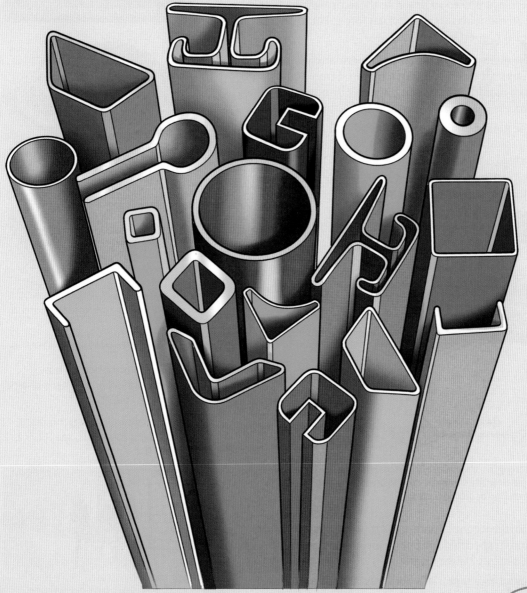

10.1 Mechanical properties

Materials behave in different ways. A designer needs to know the **properties** and **working characteristics** of materials in order to produce a successful product. One of the main factors affecting a designer's choice of material is the product's functional requirements.

Mechanical properties

Mechanical properties are used to describe those properties that indicate how a material reacts to an external force. There are six common mechanical properties:

- **Plasticity** is the ability of a material to be permanently changed in shape. An example of this is melting a material into liquid form, pouring it into a mould and allowing it to set. It now has a new shape. This process is known as **casting**.

- **Elasticity** is the ability of a material to bend and flex when a force is applied, and to return to its normal shape and size when those forces are removed – just like an elastic band.

- **Strength** is the ability to withstand force without breaking or deforming. Different forces require different types of strength to resist them, see the examples in **C**:

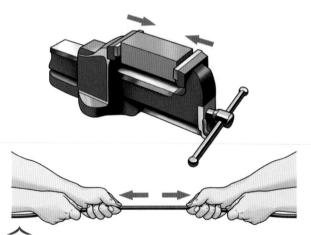

C **Compressive strength** – *the ability of a material to resist being crushed (top)* and **tensile strength** – *the ability of a material to resist stretching or pulling (bottom)*

- **Hardness** is how resistant a material is to wear. Hardwearing materials are often specified for public areas which a lot of people use every day, such as railway stations.

- **Toughness** describes a material's ability to absorb a sudden impact before breaking.

Objectives

Understand that a range of different properties affect the performance of materials.

Be able to select appropriate materials by understanding their mechanical properties.

A *A fruit jelly is produced by pouring a liquid into a mould, which then sets*

B *A silicone food mould has to be flexible in order to release the contents*

∞links

For more information on how properties affect a designer's choice of material, see 10.2.

For more about casting, moulding and forming materials, see 9.1, 9.2, 9.3 and 9.4.

D *The materials used for the floor in this public space have been chosen because they are hardwearing*

E *Both of these objects need to absorb sudden impacts without breaking*

- **Durability** is a material's resistance to wear and tear, and its resistance to corrosion and deterioration.

F *All these products have been successfully designed to be resistant to wear*

All **resistant** and **compliant materials** share these properties, but in different combinations. These different combinations determine which material a designer will specify for a product.

Key terms

Properties: how materials perform in everyday use.

Working characteristics: how a material behaves when it is shaped and formed.

Resistant materials: hard materials, such as woods, metals and plastics.

Compliant materials: materials that are flexible, such as textiles and some plastics.

Activities

1 Can you think of a material that could be successfully cast to produce a new shape?

2 Consider the chair you are sitting on. What properties would the designer need to specify when choosing a material from which to make the chair? Now consider the materials the chair is made from. Has the designer chosen the materials successfully? Explain your answer.

Activities

3 What materials would be suitable for use in public areas subject to heavy use? Explain where these materials would be used, and in what products.

4 Can you think of a situation where a tough material would be useful?

5 What functional properties would the manufacturer of soft drinks cans need to consider when specifying a suitable material? Make a list of the properties, and explain why those properties are needed.

Remember

When designing, it is important to carefully consider a product's functional requirements before choosing materials.

Summary

Mechanical properties determine which materials a designer will specify for a product.

Mechanical properties describe how a material reacts to an external force.

For many products a range of mechanical properties is needed.

Study tip

Materials have mechanical properties which make them suitable for particular products. Do you know the properties of the materials you have used in school? Drawing on your own experience of materials is a good way to answer exam questions.

Materials behave in different ways. A designer needs to know the **properties** and **working characteristics** of materials in order to produce a successful product.

Physical properties

Physical properties are used to describe those properties that indicate how a material reacts to an external force, other than mechanical. There are five common physical properties:

- **Electrical conductivity** refers to how a material resists an electric current being passed through it. If the material offers a very low resistance, it is called a conductor. Metals are good conductors, especially copper, silver and gold. Materials that offer a high resistance are called insulators. Generally, non-metal materials are good insulators, especially ceramics, glass and most plastics.

A *This soldering iron has a tip, which is a good conductor, and an insulated handle*

- **Thermal conductivity** is how heat travels through a material. A material that conducts heat very easily is said to have high thermal conductivity. Metals, especially copper and aluminium fit into this category. Materials that are poor conductors of heat, generally non-metals, are referred to as thermal insulators.

- **Magnetic properties** refer to a material that has the ability to attract or repel certain other materials. Materials with magnetic properties occur naturally. A good example of this is lodestone, which contains iron oxide. Many steels are magnetic. Magnets

Objectives

Understand that a range of different properties affect the performance of materials.

Be able to select appropriate materials by understanding their physical properties.

B *This pan is made from a material with good thermal conductivity, but has an insulating handle*

Key terms

Properties: how materials perform in everyday use.

Working characteristics: how a material behaves when it is shaped and formed.

C *Lodestone is a natural magnet*

D *An electromagnetic cat-flap*

can be made artificially by wrapping an insulated wire around an iron bar and passing a current through the wire. This is called an electromagnet.

- **Optical properties** describe how a material reacts to light, either by reflection, radiation or absorption, and will vary depending upon whether the material is opaque, translucent or transparent.
  - **Opaque materials** absorb or reflect all light and it is impossible to see through them, for example a wooden door.
  - **Translucent materials** allow some light to pass through, for example sunglasses.
  - **Transparent materials** allow light to pass through easily, for example clear glass used in windows.

- **Acoustic properties** refer to how a material reacts to sound. Some materials absorb sound, others reflect sound. Materials with good sound-absorbing properties are frequently used for products such as carpets and floor tiles. These materials are often referred to as providing sound insulation, or sound proofing. Materials with good sound-reflecting properties are used in places such as concert halls to enhance the quality of the performance.

E The lenses in sunglasses are translucent

F A soundproofing mat

Environmental properties

In addition to mechanical and physical properties, a designer must also consider the environmental properties of a material. This is a term used to describe the impact of a material's use on the environment. A material that is renewable, or can be recycled, generally has a low impact on the environment. Choosing a material that cannot be recycled or is not renewable has a bigger environmental impact. When choosing a material a designer may also have to consider other environmental issues, such as pollution (the extraction, transportation and manufacture of a material), the scarcity of a material or whether obtaining a material will involve the destruction of a habitat.

Activities

3 Make a list of the materials found in your schools workshops, and produce an environmental impact chart. Clearly show which materials are renewable, and which materials can be recycled.

4 Do you know where the materials you use in school came from? Choose one material, and then try to identify:
   a where the material originally came from
   b how it would have been transported, and
   c how it would have been manufactured into a useable form.

Activities

1 Make a list of 10 products that rely on being good conductors, and 10 products that rely on being good insulators.

2 Compasses use magnetic materials. How does a compass work?

Remember

When designing, it is important to carefully consider a product's functional requirements before choosing materials.

∞ links

To find out more about environmental issues go to 6.1 and 6.2.

Study tip

- Can you write about a material's environmental properties and, if asked, be able to select the most environmentally-friendly material from a list and match it to an appropriate product?

Summary

Physical properties determine which materials a designer will specify for a product.

Physical properties describe how a material reacts to an external force.

Designers need to consider the effects of a material's use on the environment.

Paper and card

Paper

Paper is a simple, web-like material, made from very small fibres. It is essentially a mat held together by the roughness of the fibres interlocking with each other. The fibres are made from **cellulose**, which is usually extracted from wood. Both coniferous and deciduous trees are used to provide the raw material, which is known as wood pulp. Other plant fibres can also be the source of cellulose, such as hemp, flax, cotton or even bamboo, but wood is by far the most common source.

The process of making paper was invented in China in the second century, and has remained essentially the same for around 2000 years. Tiny chips of wood are boiled in water and chemicals in order to create a mushy wood pulp. This pulp is then poured over a fine mesh. As the liquid drains away, the cellulose fibres, which are less than 1 mm in length, naturally link together to create a mat. The mat is then squeezed between a set of rollers to remove excess water, and allowed to dry.

Additives and dyes are added to the mix to change the colour of the final product and to produce different properties, textures and surface finishes. For example, toilet paper needs to absorb water quickly, whereas printing paper needs carefully controlled water absorption so that water-based inks can be used.

Paper-making takes place on a large scale. Each year, about one billion trees are cut down to make paper and paper products. Industrial-scale production of paper takes place in a paper mill.

Common forms of paper and board

Paper is available in many sizes. The 'A' sizes are the most common. A4 is twice the size of A5, A3 is twice the size of A4, A2 is twice the size of A3 and so on. Papers are also specified by weight, for example 100 grams per metre squared (gsm). This weight refers to the weight of one square metre. Anything above 200 gsm is classed as **board**.

Board is the general term for a whole range of paper-based materials, such as cardboard, carton board, mounting board and corrugated board. Board is thicker, heavier and more rigid than paper, and is made from several layers of pulp. Very thick board is made by sticking sheets of paper, or board, together in a process known as **laminating**.

Layout and tracing paper

Layout paper is a type of paper used by designers. A translucent version is known as tracing paper. These papers are used during the development stages of designing, as previous drawings can be used as an underlay. They work well with spirit-based marker pens.

Cartridge paper

This is a tough and lightly textured paper, often pale cream in colour, 100 to 135 gsm in weight. It is traditionally used for general drawing and sketching, and can also be used with watercolours.

Objectives

Be able to identify common papers and boards.

Understand the properties and uses of paper and card.

Understand the stock forms for paper and card.

Key terms

Cellulose: plant-based fibres used for paper-making.

Board: paper-based material, but thicker. Any form of paper weighing more than 200 gsm is classed as board.

Laminating: strengthening the material by building it up in layers with the same or another material.

A *Rolls of paper at the end of the paper-making process*

∞ links

Go to this website to find out more about the paper industry: **http://individual.utoronto.ca/ abdel_rahman/paper/index.html**

Cardboard

This is a heavier form of paper, often made from recycled material and available from 200 gsm upwards. Cardboard can be laminated together to create thicker boards.

Solid white board

Made from pure bleached wood pulp, strong and high in quality, this is used for book covers and expensive packaging.

Duplex board

Duplex board is made from pure wood pulp, with a bleached liner on one side. Typically 250 to 500 gsm, it is mainly used in food packaging.

Foil lined board

This is made by laminating aluminium foil to one side of any kind of board.

Corrugated board

Corrugated board is made from 'linerboard' (the flat sheet) and the 'medium' (the corrugated sheet in the middle). Cheap and often used for large cartons and boxes, it offers strength without too much weight.

B *Different types of paper and board*

∞ links

For information about paper and board used in packaging, see 4.2.

To learn more about how paper and board are combined with other materials to improve their properties, see 11.1.

Remember

Paper and board can be produced with a wide range of different properties, textures and surface finishes for different applications.

Study tip

- Learn about paper and board as these are compulsory materials for every student.
- Can you match the different types of paper and board to specific products?

Activities

1. Paper is regarded as a sustainable product. What do you think this means?

2. For each type of paper and board, give an example of how it might be used in a product.

Summary

Paper and board is generally produced on an industrial scale, and is usually specified by size and weight.

Paper-based materials have a wide range of uses, and can be laminated with other materials for specific purposes.

Wood (or timber) has been used as a structural and decorative material for thousands of years. Natural wood is sawn directly from trees, while manufactured boards are commercially made by gluing pieces of wood together.

Designers make choices about which timber-based material they are going to use by considering whether the properties suit the needs of their design. They may also consider environmental issues. Wood-based materials are more **sustainable** than other materials, and are commonly identified by the Rainforest Foundation logo.

Natural wood

There are two basic types of tree: **hardwood** and **softwood**. Hardwoods are generally deciduous, while softwoods are generally coniferous (often called evergreen). The size of natural timber is determined by the size of the tree. All natural woods are **seasoned**. Approximately 80 per cent of the wood used in the UK comes from other countries.

Hardwood trees grow more slowly than softwoods. Examples of hardwood trees grown in the UK include oak, ash, beech, sycamore and willow. Imported tropical hardwoods include teak and mahogany.

B Oak is a hardwood

C Scots pine is a softwood

Softwood, which grows quickly, is often managed as a **sustainable** resource. There are a smaller number of useable softwoods than useable hardwoods. Some softwoods (larch, spruce, and Scots pine) are grown in the UK, but approximately 90 per cent of the softwood used in the UK is supplied by countries such as Norway and Sweden.

Designers choose natural wood for its variations in colour and pattern: as sourced from individual trees, no two sections will be the same.

Manufactured wood

Manufactured, or man-made, wood is board produced using industrial production techniques. It consists of gluing together wood layers or wood fibres. Manufactured boards are usually made in very large sheets.

A The FSC symbol – used on wood from sustainable sources

Key terms

Hardwood: a natural material generally sourced from a deciduous tree.

Sustainable: capable of being produced with minimal long-term effect on the environment.

Softwood: a natural material generally sourced from a coniferous tree.

Seasoned: all natural wood is seasoned to remove moisture. The process of seasoning involves drying in a controlled way to prevent twisting and warping.

Manufactured boards: timber-based products produced by an industrial process involving gluing smaller pieces together.

∞ links

To find out more about sustainability see 6.1, or visit **www.bbc.co.uk** and search for: Bloom sustainable wood.

Designers choose **manufactured boards** when they require consistency in strength, workability and texture. Their plain appearance is often disguised by more decorative material. Common manufactured boards include:

MDF – a material composed of fine wood dust and resin pressed into a board

plywood – made from thin layers of wood, glued at right angles to each other

chipboard – made from small chips of wood glued together with resin and compressed into sheets

blockboard – made from a core of softwood strips glued together and covered with a sheet of plywood on either side

hardboard – made from wood fibres that have been pulped, then compressed.

MDF

plywood

chipboard

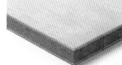
blockboard

hardboard

D *Manufactured boards*

Common forms of wood

Hardwoods are usually sold by the cubic metre, then rough sawn to the size the customer specifies. Hardwoods are also machined into a variety of standard sections called mouldings, for example dowel and quadrant beading.

Softwoods are usually supplied in standard sections, rough sawn or planed smooth. Sizes are often confusing because once planed the section will be smaller (for example a 50 x 50 mm section is likely to be about 45 x 45 mm once planed).

The majority of manufactured boards are supplied in Imperial sizes (feet and inches). The most common size is 8 x 4 feet (2240 x 1220 mm). Boards are available in a variety of thicknesses, such as 3 mm, 6 mm and 9 mm. Not all boards are available in every thickness.

⬭ links

To find out more about manipulating and combining timber-based products, see 11.1.

Remember

Wood is a sustainable product.

Study tip

- If wood is the main material you have studied, can you explain the difference between hard and soft woods?
- Can you recognise the properties of some common woods for use in product manufacture?

Activities

1. A designer of children's toys would like to make a new product from a sustainable natural wood. Name a suitable timber, giving the reasons for your choice.

2. Timber is often referred to as a sustainable material. What does this mean?

3. Manufactured boards are often used in places where there are variations in temperature. Explain why this is the case, giving examples.

Summary

Natural timber comes from hardwood and softwood trees, and is often chosen because of the variation in colour and pattern it provides.

Manufactured boards provide consistency in strength and texture.

10.5 Metals

Metals have been used as a material for two thousand years. Apart from gold, which is obtained in a useable form, all other metals are found as ores, which are produced by mining. Metals are extracted from ores in a large-scale industrial process. Metals are non-renewable, so **recycling** them is important.

Metals are usually separated into two groups, ferrous and non-ferrous, each of which contains pure metals and alloys. Designers make choices about which metals they will use by considering their various properties.

Ferrous metals

Ferrous metals contain iron; they include cast iron, mild steel, high carbon steel and stainless steel. Steel, one of the most important metals we use, is produced by mixing iron with a small amount of carbon. The amount of carbon present determines its strength. Ferrous metals are prone to a form of corrosion called rusting.

A | Metal gear wheels

B | Gold rings

Non-ferrous metals

Non-ferrous metals do not contain iron, and include aluminium, copper, lead, zinc, tin, gold and silver. In general, the rarer the metal, the more expensive it is. However, some metals are more difficult to produce, which also adds to their cost.

Alloys and alloying

Pure metals have a limited range of properties. By combining two or more metals to produce an **alloy**, it is possible to produce a material that has a better range of properties and characteristics, thus making metals more suitable for specific tasks. The majority of metals we use are alloys.

By alloying metals it is possible to:

- change the melting point
- change the colour

C | A steel rollercoaster

- change electrical/thermal properties
- increase strength, hardness and ductility.

Brass is a non-ferrous alloy. It is made from 65 per cent copper and 35 per cent zinc, and is harder and stronger than either of these metals in their pure form. Stainless steel is a ferrous alloy and is tough and resistant to wear and corrosion. It is made from 74 per cent iron, 18 per cent chromium and 8 per cent nickel.

Common forms of metal

Metals are available in sheet form and sections. Sheets are usually sold in imperial sizes, like timber, but in metric thicknesses such as 1 mm and 2 mm. Many specialised sections such as hexagonal, angled and channels are available in both ferrous and non-ferrous metals. Sections are often stiffer than the material in solid form, but allow less material to be used, saving weight and reducing costs. Assuming that both have the same cross-sectional area, a tube section is stiffer than a solid rod made from the same material. Common sections include: 'T' section, 'L' section, square section (solid and tube), round section (solid and tube) and channel section.

D *Various metal sections*

Key terms

Recycling: a way of re-processing and re-using a material.

Ferrous metal: a metal containing iron.

Non-ferrous metal: a metal that does not contain iron.

Alloy: a metal produced by combining two or more metals.

⚭ links

To find out more about recycling go to 6.1.

For more information on metals recycling visit www.wasteonline.org.uk

Find out more at
Check out 11.2 for information on plating and galvanising metals.

Remember

Metals are a non-renewable material, so it is important that they are recycled.

Study tip

- If metal is the main material you have studied, can you explain the difference between ferrous and non-ferrous metals?

- Check you can recognise the properties of some common metals for use in product manufacture.

Activities

1 Steel is the most commonly used material for the manufacture of car bodies. Why is this? Aluminium is sometimes used for the same purpose. What are the advantages and disadvantages of using aluminium instead of steel?

2 Explain why it is important to recycle metals.

Summary

Metals can be divided into those that contain iron (ferrous) and those that do not (non-ferrous).

Two or more metals can be combined to create 'alloys'.

10.6 Plastics

Unlike wood and metal-based materials, the mainly synthetic materials we refer to as plastics are relatively new. Since prehistoric times, people have used natural polymers such as horn, waxes and bitumens; however, it was not until the 19th century that the first synthetic plastics were developed. Research and development still continues, with new materials and applications being found almost daily. The raw material for most plastic products is non-renewable, so **recycling** is important.

Plastics are now the most widely used materials in commercial production. The ease of moulding plastics into complex shapes enables the designer to develop innovative products. Plastics can be created from both natural and synthetic sources.

Natural plastics

Natural plastics include materials such as amber, which is fossilised tree resin, and latex, which is a form of rubber.

Synthetic plastics

Synthetic plastics are by far the most common form of plastic used today. Also known as polymers, they are produced industrially from carbon-based materials, such as coal, oil and gas, through a process known as **polymerisation**. Polymerisation has two stages: firstly, monomer molecules are created from a raw material; secondly, these are then joined to form long chains of molecules called polymers.

Thermoplastics

Thermoplastics are formed from long polymer chains, joined loosely together. Heating weakens the join and softens the plastic, allowing the heat-softened material to be reshaped. The thermoplastic hardens when cooled, but can be reshaped if reheated. Common thermoplastic materials include: high-density polythene (HDPE), low-density polythene (HDPE), polypropylene, high-impact polystyrene (HIPS), nylon, polyvinyl chloride (PVC) and acrylic.

Thermosetting plastic

Thermosetting plastics are formed when a polymer, usually in the form of a resin or powder, is mixed with a catalyst to produce a chemical change, resulting in a permanently rigid material. Thermosetting plastics, once moulded, cannot be re-formed. This is because the polymer chains are so tightly linked, no amount of re-heating or manipulation will weaken them.

Reinforced plastic

Thermoplastic and thermosetting plastics can be reinforced by the addition of fibres, such as glass or carbon fibre.

Objectives

Be able to identify common plastics.

Understand the differences between thermoplastics and thermosetting plastics.

Understand the stock forms for plastic materials.

links

For more about recycling, see 6.1.

For an information sheet about plastic recycling see www.wasteonline.org.uk

A *An amber pendant*

links

To find out more about plastics go to:
www.bpf.co.uk/Plastipedia_a_Guide_to_Plastics.aspx

www.howstuffworks.com

See 11.2 for information on the new biodegradable plastics.

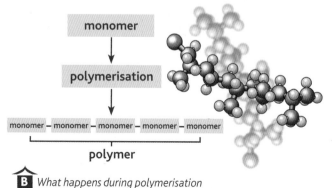

monomer

↓

polymerisation

↓

monomer – monomer – monomer – monomer – monomer

polymer

B *What happens during polymerisation*

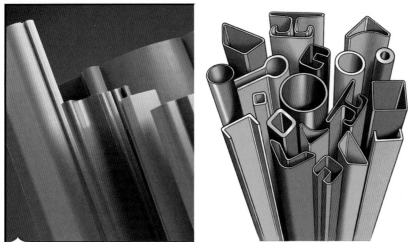

C *Extruded plastic products. The graphic shows the range of profiles available*

Key terms

Recycling: a way of re-processing and re-using a material.

Synthetic: a manufactured material.

Polymerisation: the process of creating a synthetic plastic.

Thermoplastics: plastic materials that can be remoulded with heat.

Thermosetting plastic: a plastic material that, once moulded, cannot be reshaped.

D *Products made from thermoplastics*

F *A glass reinforced plastic- (GRP) bodied sports car*

E *Products made from thermosetting plastics*

Common forms of plastic

Most thermoplastics are available in different thickness sheets, such as 1 mm and 2 mm. Some are also available as granules (for injection moulding), powders (for processes such as dip coating), rods and blocks. Thermosetting plastics are most commonly found in resin or powder form.

Remember

It is important to recycle plastics as the raw material for most plastics is non-renewable.

Study tip

- If plastic is the main material you have studied, can you explain the difference between thermoplastics and thermosetting plastics?
- Can you recognise the properties of some common plastics for use in product manufacture?
- Would you be able to draw diagrams to explain the different manufacturing processes used to turn them into products?

Activities

1   Explain why a designer would specify a thermosetting plastic for a frying pan handle rather than a thermoplastic.

2   A manufacturer of school chairs has decided to use high-density polythene for the seat. What three properties does this plastic have that makes it an ideal choice? Give your reason for each choice.

Summary

Most modern plastics are synthetic.

Plastics can be classified as thermoplastics or thermo-setting plastics.

Plastics can be combined with other materials to reinforce their structure.

10.7 Ceramics

We tend to think of ceramics in terms of crockery (plates, cups and saucers) or decorative products (vases, tiles and figurines). However, clay-based ceramic products are widely used in engineering products, including house bricks, engineering bricks, the heat-resistant tiles used on the NASA space shuttle and the brakes used on high performance cars.

Ceramic materials are inorganic (composed of minerals rather than living material) and non-metallic. The working characteristics of ceramics vary depending on the amount of moisture in the material. Moist or wet ceramics at room temperature are flexible; dry ceramics are brittle or crumbly. When heated to high temperatures in a kiln, ceramics become very hard and cannot be recycled with water.

In many parts of the world, clay is dug straight from the ground and used for making a variety of pottery items by hand. Modern clays may be dried, crushed and mixed with other minerals and exact quantities of water in order to obtain ceramic materials that are consistent in quality and plasticity. Common ceramic materials include **earthenware**, **stoneware** and **porcelain**.

Earthenware

Earthenware is fired at a low temperature (1 200 °C), after which it is porous. It includes 'terracotta' clays (from the Italian word meaning 'baked earth'), which are usually red in colour. Terracotta is the cheapest clay, found in large deposits in many parts of the world. It is used for tiles, bricks and domestic ware.

Stoneware

Stoneware is fired at a higher temperature than earthenware (1 200–1 280 °C). Once fired it produces a non-porous (or vitreous), glass-like ceramic (with a composition similar to stone) and is used in both the arts and industry. Domestic stoneware pots are heavier and more durable than porcelain or earthenware.

Porcelain

Porcelain, also known as china clay or kaolin, is fired at high temperatures (1 400 °C), and can be opaque or translucent, depending on the make-up of the clay body. Porcelain is difficult to handle and is used mainly by expert potters. Once fired, it is tough and has a smooth, shiny surface.

C *Fine porcelain china*

Objectives

Be able to identify common clays and related products.

Understand the properties of ceramics in relation to domestic pottery and the electrical industry.

Understand the stock forms of clay.

∞ links

To find out about designers who work with ceramic materials, take a look 'Ceramic Design' in the directory at:
www.dexigner.com

A *Earthenware pots*

B *Stoneware mug*

∞ links

To find out more about modern uses of ceramic-based materials, click on 'Applications' at:
www.dynacer.com

Glazing

Fired ceramics usually require a glaze to seal the surface. Chemicals are often added at this stage of the process in order to colour or decorate the finished product. There are numerous recipes for the composition of a **glaze**, and the results can vary even when a recipe has been followed carefully. The application of a glaze is followed by a second firing.

Technical ceramics

Increasingly, ceramics have become 'designer' material, the properties of which can be changed by combining materials in much the same way as metal alloys are created for specific applications. A metal-oxide based ceramic material, created by combining ceramics with a substance called Beryllia, has a melting point of 2 350 °C and is used to manufacture crucibles for nuclear reactors. An aluminium-based metal-oxide is used to produce bulletproof vests.

Ceramics are being hailed as the materials of the future, for example in the automotive industry. Because they are heat-resistant, ceramic car engines could operate at very high temperatures, requiring no lubricants and, as petrol burns more efficiently at higher temperatures, energy wastage could be reduced. Ceramic engines would not need cooling systems, thus reducing their weight and leading to even greater fuel efficiency.

Common forms of ceramics

Clay is most commonly sold by weight (for example, 25 kg). Clay in moist form is called 'body' clay. Liquid clays (often used for decoration) are known as slip. The ingredients used for colouring and glazing ceramics are usually sold as dry powders.

Other inorganic based materials

Other materials produced from inorganic sources include:

- glass (produced from silica, also called sand)
- cement (produced from limestone)
- plaster of Paris (produced from a very soft mineral called 'gypsum', large deposits of which were found near Paris).

D *Ceramic bearings*

Remember

Ceramic-based materials are widely used in engineering applications.

Study tip

- If ceramics is the main material area you have studied, you should show your knowledge by selecting ceramic products, components and materials in the materials questions.
- Can you name three different types of ceramic materials and explain how they are used in new products?
- You should also be able to understand and explain the related manufacturing processes used to turn them into products.

Activities

1. Earthenware is porous and stoneware is vitreous. What practical uses could each type of clay be used for?

2. Porcelain has many industrial uses. Can you name some examples of its application?

Summary

Ceramics are produced from clay, a natural material.

Common ceramics include earthenware, stoneware and porcelain.

The properties of ceramics can be changed by adding other materials to produce 'technical ceramics'.

A textile is a flexible material, or fabric, comprised of a network of fibres referred to as thread or yarn. Until recently, all textiles were made from **natural fibres**, but with the growth of the plastics industry in the 20th century, these were supplemented by **regenerated** and **synthetic fibres**.

Fibres

Fibres are fine, hair-like structures which can be natural, regenerated or synthetic. Short fibres from sources such as cotton or animal hair are called 'staple' fibres. Long continuous fibres, such as silk, are known as 'filament' fibres.

Natural fibres

Natural fibres can be subdivided into **animal**- or **plant**-based fibres. All natural fibres have to be processed to make them suitable for use as yarns and fabrics.

Animal fibres are commonly made from hair or fur. Wool refers to the hair of sheep or goats. Silk is an animal-based fibre made from the fibres of the cocoon of the Chinese silkworm, spun into a smooth, shiny fabric.

Cotton, the most important plant-based fibre, comes from the fine hairs on the seeds in a ripe seed pod of the cotton plant. Linen is produced from the stalks of the flax plant.

Regenerated fibres

Regenerated fibres are made from cellulose-based fibres of plant origin (such as wood pulp). The classification of the fibre relates to the chemical solvent system used to extract the fibre, so regenerated fibres are part natural and part manufactured, or man-made. Regenerated fibres, such as viscose, rayon, acetate, triacetate, modal and Tencel, are widely used in clothing. Different finishes make them smooth, shiny or textured.

Synthetic fibres

Synthetic fibres are made from chemicals obtained from oil, coal or other petrol-based chemicals. Fibres are drawn out into long threads or filaments, usually by spinning, or by melting the material and then forcing it through the holes of a perforated plate. Synthetic fibres can be developed with a range of properties, and can be classed as thermoplastics. Common synthetic fibres include polyester, nylon and acrylic.

Fabrics

Fabrics can be knitted, woven or non-woven. Fibres for knitted and woven textiles are spun or twisted together to form one continuous fibre, usually called yarn, which is then used to make fabric. During the spinning process, the fibres are twisted either anticlockwise (S-twist) or

A *Knitted and woven fabrics*

B *Products made by felting*

clockwise (Z-twist). For more complex yarns, S-twist and Z-twist yarns are combined in equal amounts to prevent distortion. After spinning, the yarns are knitted or woven into large sheets.

Woven textiles are made on machines called looms, while knitted fabrics are made by forming locks that intertwine one or several fibres. For mass production, knitted textiles are made by machines with many hooked needles. Some machines have up to 2500 needles, forming 3 000 000 loops per minute.

Non-woven fabrics are produced by bonding or felting. Bonding is the process of joining synthetic fibres together with resin. When the resins have set, the fabric is heated to make it stronger. In some fabrics, the fibres are pressed through two hot metallic rolls to join them. The fibres melt and form strong bonds, which keep their resistance after cooling. Felting produces a non-woven fabric by matting together animal hair or wool fibres, using moisture, heat and pressure.

Technical textiles

The development of synthetic fibres has transformed textiles. Modern textiles are made from combinations of materials and can be 'designed' for specific tasks. Technical textiles are manufactured mainly for their performance and functional properties, for example:

- Clothing textiles. Performance materials such as GORE-TEX are waterproof, windproof and highly breathable.
- Automotive textiles. The internal structure of a tyre is made-up from textile fibres including cotton, nylon and polyester.
- Aerospace textiles. Woven fabric structures form part of composite materials used in aircraft manufacture.

C Products designed and manufactured from technical textiles

Summary

There are three groups of fibres: natural, regenerated and synthetic.

There are three types of fabric: knitted, woven and non-woven.

Synthetic fibres have led to the development of 'technical textiles'.

∞ links

To find out more about textiles, and how their properties can be combined or manipulated, go to 11.3.

Remember

Synthetic fibres are manufactured (man-made) and use up natural resources such as coal and oil.

Study tip

- Can you match types of fibres to products, and do you know the stock forms for the various textiles materials?

Activities

1 Plastics are commonly used as a material from which to manufacture clothes. Why do you think some clothing manufacturers use plastic-based synthetic materials rather than natural fibres?

2 Consider a product made from a synthetic material. Specify the material used and explain why it is suitable for its intended purpose.

3 Some people prefer to wear cotton next to their skin, rather than synthetic fabrics. Why is this?

4 NASA spacesuits are made from a mixture of textile-based materials. Find out what fabrics they are made from, and how they are made. Why is a spacesuit white?

To be successful, a food product has to appeal to its target market. Tastes change over time, and there are many reasons why people buy different foods today compared with 20 years ago. However, whatever the product, a food designer must consider the **nutritional value** of the food components, known as ingredients, which make up the finished product. Food provides **nutrients** to help the body work properly. As no food on its own provides these nutrients in the amounts needed by the body, a mixture of foods has to be eaten for a balanced diet.

Food groups

Food materials can be divided into five main groups:

- bread, rice, potatoes, pasta and other starchy foods
- fruit and vegetables
- milk and dairy foods
- meat, fish, eggs, beans and other non-dairy sources of protein
- foods and drinks high in fat and/or sugar.

Choosing foods from the first four groups every day will help ensure that the body receives good nutrition. It needs a wide range of nutrients to remain healthy and function properly. Choosing different foods from within each group adds to the range of nutrients consumed. Foods in the fifth group (those containing fat and sugar) are not essential to a healthy diet, but increase choice.

A balanced diet includes:

- proteins, which are required for body growth and repair
- fats, which are needed for energy and to keep the body healthy
- carbohydrates, which provide energy, and work with proteins to aid growth and repair
- vitamins, to prevent illness and control the release of energy in the body
- minerals, which help in building the body and controlling how it works.

To help individuals achieve a balanced diet, the Food Standards Agency has developed the 'eatwell plate' to show how much should be eaten from each food group daily.

Objectives

Be able to classify food into common groups.

Understand the importance of a balanced diet.

Understand that food components are available in a variety of forms, including standard components.

Key terms

Nutrition: how our bodies take in and use food.

Nutritional value: the amount of energy that a food gives you when you eat it. It is measured in either calories (kcal) or joules (kJ).

Nutrients: substances from food that give us energy, help repair body tissues, and regulate body functions. There are six different types of nutrients: carbohydrates, fats, proteins, vitamins, minerals and water.

The eatwell plate

Use the eatwell plate to help you get the balance right. It shows how much of what you eat should come from each food group.

A The 'eatwell plate'

The 'eatwell plate' is based on the Government's guidelines for a healthy diet:

- Enjoy your food.
- Eat a variety of different foods.
- Eat the right amount to be a healthy weight.
- Eat plenty of foods rich in starch and fibre.
- Eat plenty of fruit and vegetables.
- Don't eat too many foods that contain a lot of fat.
- Don't have sugary foods and drinks too often.

Common forms of food

Most food ingredients are available fresh, and supplied by weight or volume, for example grams and litres. Many commonly available food products have been preserved to extend their shelf-life. Preservation destroys the bacteria present in food, or prevents it from reproducing.

Traffic light labelling

Some supermarkets and food manufacturers have developed a 'traffic light' scheme to help consumers be aware of the levels of fat, sugar and salt in a product. A red 'light' indicates that the food is high in something we should be cutting down on, 'amber' is okay most of the time and 'green' shows that the food is low in a specific nutrient. The more 'green lights', the healthier the product.

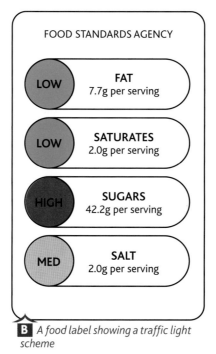

FOOD STANDARDS AGENCY

LOW — **FAT** 7.7g per serving

LOW — **SATURATES** 2.0g per serving

HIGH — **SUGARS** 42.2g per serving

MED — **SALT** 2.0g per serving

B *A food label showing a traffic light scheme*

links

Look at 9.11 for more information about the methods used to condition and preserve food.

For information on the 'eatwell plate', go to:
www.eatwell.gov.uk
The same website gives you a link to the traffic light labelling scheme.

Remember

A healthy diet contains: plenty of fruit and vegetables; starchy foods such as wholegrain bread, pasta and rice; and is low in fat (especially saturated fat), salt and sugar.

Study tip

- Can you match food groups to food products?
- Do you know the stock sizes for the various food materials?

Activities

1. Make a list of the food you have eaten during the past 24 hours under the headings of:
   - protein
   - carbohydrates
   - fats.
   - vitamins
   - minerals

2. State three reasons why you might need to know the ingredients in a food product.

3. Food is often preserved. Why do you think this is necessary? Describe two different methods of preserving food, naming the examples you have chosen.

Summary

Food can be divided into five main groups.

A balanced diet helps a body remain healthy.

Preserving food can extend its shelf-life.

Food labelling helps consumers choose healthier products.

10.10 Electronic control components

At the heart of every electronic product is a **printed circuit board** that contains individual (or discrete) **components** all working together. Common components include resistors, capacitors and transistors. All these are attached to a printed circuit board by **soldering**.

Discrete components

- Resistors, in fixed or variable values, control the flow of electric current in a circuit. Resistance is measured in ohms.
- Capacitors store electrical charge, and are used when a time delay is required or when a quick burst of current is needed, as for a camera flash.
- Transistors are used as electronic switches where a small current is used to switch on a larger current. Sensitive to small currents, they are used in sensing circuits.

Integrated circuits

Many discrete components have now been replaced by integrated circuits (ICs or 'chips'). These are miniature circuits with many components, all etched onto a single piece of silicon and encapsulated within a protective plastic package. Programmable chips are known as programmable integrated circuits (PICs).

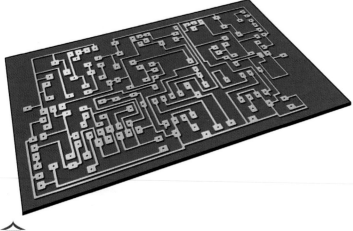

A *A printed circuit board (PCB)*

Electronic systems

Electronic circuits are described as systems. Each system is made up of: input, process and output.

B *A system diagram*

C *Switches*

Inputs

Inputs activate a circuit and include:

- switches (the main input in most circuits), including push to make, push to break, toggle and slide
- light dependent resistors (LDRs), which convert light energy to electrical energy
- thermistors, which convert heat energy to electrical energy
- microphones, which convert sound energy to electrical energy.

Outputs

Outputs are the result of the input being processed and include:

- bulbs, which convert electrical energy into light and heat
- light emitting diodes (LEDs), which convert electrical energy into light. Because of their low consumption of power and long life, LEDs are replacing bulbs in many applications
- motors, which convert electrical energy into rotational movement
- buzzers, often found in alarm circuits, which convert electrical energy into sound
- solenoids (or electromagnetic devices), which convert electrical energy into small linear movements
- bells, which are also electromagnetic devices – the movement generated by the electromagnet causes a small hammer to strike a bell
- loudspeakers, which change electrical pulses into sound – the volume and pitch depends upon the pulses received.

Microcontrollers and microprocessors

Many electronic products are based around programmable devices. Microprocessors are used as multifunction programmable devices; microcontrollers are used for specific functions. Using a microprocessor to control complex equipment means that one device can control everything. But if the microprocessor malfunctions, all the systems could be affected and the replacement cost may be high. If microcontrollers are used instead, a damaged one can be easily replaced.

Summary

Electronic circuits are usually made by soldering individual components onto a printed circuit board.

Electronic systems consist of an input, process and output.

Many electronic products contain programmable microcontrollers or microprocessors.

Remember

An integrated circuit (IC) is also called a chip; it can contain dozens, hundreds or millions of electronic components.

∞ links

If you would like to discover more about electronics, go to the online resources at:
www.doctronics.co.uk

Activities

1. PICs are increasingly being used instead of dedicated circuit boards. Explain why this is the case.

2. Many products now use LEDs instead of incandescent and fluorescent light sources. Explain their advantages for the manufacturer and the consumer, using examples to illustrate your answer.

3. Explain the advantages, for the environment and the consumer, of using rechargeable instead of disposable batteries.

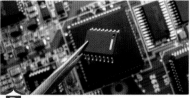

D *A computer microprocessor*

Mechanisms generate force and movement within products to make tasks easier. Mechanisms can be small and precise (such as the internal workings of a watch) or large and powerful (such as a crane or mechanical digger).

Mechanical systems

Mechanisms are mechanical systems that convert movement and transmit power. They have inputs and outputs but, unlike electronics, mechanical components are not limited to one type of action. A crank can be an input or an output.

Types of movement

These are:

 linear – moves in one direction
■ reciprocating – moves backwards and forwards
■ rotary – turns in a circle
■ oscillating – swings in alternate directions.

A *Linear motion*

B *Oscillating motion*

C *Rotary motion*

D *Reciprocating motion*

Types of mechanism

Most mechanisms consist of combinations of basic components: levers, linkages, pulleys, chain and sprockets, gears, cams and cranks. Mechanisms can also include pneumatics and hydraulics.

A lever pivots around a fixed point, a **fulcrum** or **pivot**. Many hand tool designs are based on levers, which are used to increase force.

Linkages

Linkages are used to transfer motion between mechanisms. Simple linkages often use levers to change direction or multiply force.

E *A pair of scissors, which consists of two levers pivoting around a fixed point*

F *A chain and sprocket mechanism found on a cycle*

Pulleys

Pulleys are used to change the direction or speed of rotary motion, or make lifting heavy loads easier. Pulleys are usually connected by belts, but can also be used with cables and ropes to lift loads. In many pulley-based machines (for example sewing machines), one pulley is driven by an electric motor.

Chain and sprockets

A chain and sprocket mechanism (commonly found on cycles) works in a similar way to pulleys, but with less risk of slippage.

Gears

Gears are wheels with teeth cut into their outer edge. Unlike pulleys and sprockets, they do not need a belt or chain between them, relying on the teeth of the gears to engage, or 'mesh'. Gears are used in pairs and groups to transmit rotary motion, change direction or speed. Special gears can be used to transmit rotary motion through a right angle.

Cams and cranks

Cams are wheels with a shaped profile. A follower runs around the edge of the cam and produces reciprocating motion. The motion of the follower can be modified by using a different shaped cam.

Cranks are shafts that are stepped along their length. By connecting a rod, a rotary motion will produce a reciprocating motion. This can work in reverse: when driven by a piston, the crank will produce a rotary motion.

Pneumatics and hydraulics

Pneumatics (compressed air) and hydraulics (compressed fluid) are technologies that often interface with mechanical systems.

- Pneumatic systems use the potential energy stored in compressed air. By controlling the release of air, this energy can be turned into movement. A dentist's drill is powered by pneumatics. It is lighter and faster than its electric equivalent.

- Hydraulic fluids cannot be compressed but, by forcing fluid from a small cylinder to a larger cylinder, it is possible to increase the output force. Combining this force magnification with levers creates a very efficient machine. This technique is employed to control the arms on mechanical diggers.

G *Spur gears*

H *Bevel gears*

I *Crank shaft and pistons*

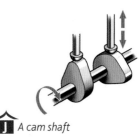

J *A cam shaft*

∞ links

If you would like to discover more about mechanisms, go to the 'Mechanisms' link at **www.flying-pig.co.uk**

> **Remember**
> Mechanisms can help us to move heavy loads by magnifying a smaller force.

Summary

Mechanical systems convert movement and transmit power.

There are four types of motion: linear, reciprocating, rotary and oscillating.

Mechanisms consist of a range of basic components.

Mechanisms can be interfaced with other technologies.

11.1 Paper- and timber-based products

Paper- and board-based products

Paper and board are very versatile materials and, combined with other materials, can be found in many everyday products. Paper-based packaging materials are often the most economical form of packaging for the food industry. Cardboard cartons are popular for packaging as they are economical, collapsible, printable, can be shaped easily, can have dispensing and/or resealing features and windows for viewing the product, and can be used in multipacks. Corrugated packaging in the form of boxes provides protection for a variety of products, especially during transportation. Lined board is used in the manufacture of **aseptic** cartons, which consist of thin layers of paper-based board, aluminium foil and polythene film **laminated** together to combine the best properties of each individual material into a single packaging material.

A popular aseptic package is the **Brik Pak**. Its **composite** structure usually contains cardboard (70%), polythene film (24%) and aluminium foil (6%). Cardboard is a mechanically stable material with good heat resistance. It also provides good light protection and a printable surface. Aluminium foil is an excellent gas, water and light barrier (particularly when laminated between two plastic layers) and it is thermally stable. The combination of these properties has made paper/foil/plastic laminations popular in aseptic packaging, which is used extensively for products such as fruit juices, milk and soups.

The one major drawback of combined packaging materials is that recycling them is difficult because it is hard to separate the various materials.

A Examples of aseptic packaging used for food products

Timber-based products

Natural wood is also a very versatile, and popular, material, but its use is often limited by cost, the fact that it can become unstable in some situations and has limitations in size. As a result, manufactured products are constantly being developed to address these issues while retaining the beauty of natural wood.

Objectives

Understand that materials can be combined and processed to create more useful, or desirable, materials.

Understand that paper-based boards can be laminated to other materials.

Understand that the composition of timber-based products can be adjusted to create different properties for specific purposes.

⊂⊃links

To find out more about paper-based products, see 10.3.

To find out more about manufactured boards, see 10.4.

Key terms

Laminated: the process by which layers of material have been joined together.

Brik Pak: a term used to refer to a common form of aseptic packaging.

Composite: made up of more than one thing.

Laminate: a material formed by bonding two or more layers together; the layers are usually different materials.

⊂⊃links

To find out more about aseptic packaging, go to:

www.tetrapak.com then click on Products and Services, followed by Packages **www.aseptic.org**

Manufactured boards are made in large sheets, sometimes from waste products, and are usually very stable. However, although they have many advantages over natural wood, manufactured boards are generally unattractive. To solve this problem, manufactured boards are often produced with a decorative layer called a veneer. A veneer is the term given to thin layers of wood (typically 0.8 mm thick) cut from the trunk of a tree. The trees are usually hardwood trees as their wood has a more interesting appearance.

The use of veneers can reduce the cost of a product because manufactured boards are considerably cheaper than natural woods, which are generally too expensive for mass-market products. IKEA is one example of a company that makes extensive use of this type of material.

Manufactured boards can also be covered in a plastic **laminate**, and in this form are commonly used for kitchen worktops and laminate flooring. Kitchen worktops are made from layers of printed paper in a plastic resin. The paper gives a decorative finish and the resin provides a hardwearing, heat-resistant coating.

B *Examples of veneered and laminated furniture and worktops*

Activities

1. Laminated board is widely used in the food packaging industry. Why do you think this is? Give one example of a food product which makes use of laminated board for its packaging, and state the type of board used.

2. Alvar Aalto is a Finish designer, famous for producing furniture from laminated wood. Produce a short PowerPoint presentation about his designs, and how they are made.

3. Disassemble a Brik Pak product you are familiar with and try to identify the various component parts.

4. By following the web links at the foot of page 122, can you suggest why manufacturers use Brik Paks instead of conventional packaging?

Summary

Paper and board can be combined with other materials to produce a range of packaging materials.

Composite packaging materials are widely used in the food industry.

Manufactured boards can be combined with natural woods to create a product that overcomes the weaknesses of the individual materials.

Metal- and plastic-based products and composite materials

Metal-based materials

Non-ferrous metals have a protective natural oxide layer to prevent **corrosion**. Ferrous metals, with the exception of stainless steels, corrode by a process known as rusting. To prevent this, there is a range of processes by which a protective coating can be applied.

Electroplating

This electrochemical process is used to coat a base metal with other metals. The coating material, in addition to preventing **corrosion**, gives the base metal a more attractive and durable finish. Bathroom taps, coated with chromium, are produced using this process.

Dipping

Metals can also be coated by dipping them into another metal.

Tin plating

Tin plating is a process in which steel sheets are passed through tanks of molten tin at a high temperature. Materials produced in this way are used to manufacture cans for the food industry.

Zinc plating

Zinc plating is the process of dipping steel into vats of molten zinc. This process is also known as **galvanising**, and is widely used in the production of car body shells.

Enamelling

Enamelling is the process of coating ferrous and non-ferrous metals with a layer of finely ground glass. This involves heating the metal to a high temperature so that the enamel fuses together to provide a hard coating. When used with steels, enamel provides a layer of heat- and scratch-resistant material, and is often used as a coating for cookers. Fine enamels are used with gold and silver as a means of applying colour to jewellery.

Plastic coating

Plastic coating is the process of applying a thermoplastic to the surface of metal items to provide long-term corrosion, impact and chemical resistance. It also gives an attractive decorative finish.

A *A chrome plated tap*

B *An enamelled silver ring*

Objectives

Understand that materials can be combined and processed to create more useful, or desirable, materials.

Understand that because most plastics are 'synthetic', their composition can be adjusted to create different properties for different purposes.

Understand that, by combining materials from different material areas, composite materials can be developed to provide outcomes with different properties for different purposes.

Key terms

Corrosion: the deterioration of a metal, usually caused by a chemical reaction, and often linked to its environment.

Galvanising: the process of coating steel with a thin layer of zinc to prevent corrosion.

∞ links

Processes such as plating, galvanising and enamelling are also covered in 9.12.

To find out more about metals, see 10.5.

To find out more about plastics, see 10.6.

To read more about carbon fibre, see 11.5.

To find out more about bio-packaging, go to:

www.londonbiopackaging.com

Plastic-based materials

Plastics are largely synthetic or manufactured materials so can be easily processed to provide a wide range of different properties. Fillers, such as sawdust, crushed quartz and limestone, can be added to give the material bulk, while other additives can be used to condition or change the properties of the material. Plasticisers can be added to improve the flow properties when injection moulding plastics, while flame-retardants are often used to produce a plastic which is resistant to combustion. Additives can also be introduced to make the plastic more biodegradable.

Biodegradable plastics are made entirely from renewable raw materials, such as wheat and corn, as well as potatoes (a very recent development). Bio-based plastics, as these materials are known, are widely used in the food industry as a packaging material (bio-packaging).

Composite materials

Composite materials are made from two or more different materials, with each material still retaining its own identity in the new material. The advantage of this is that the properties of each can be enhanced and utilised. Common composite materials include: glass-reinforced plastic (GRP) and reinforced concrete.

Glass-reinforced plastic

Glass-reinforced plastic (GRP), also known as fibreglass, is a composite material made from spun glass fibres and polyester resin. The final material is tough, rigid and lightweight. GRP is often used to produce boat hulls and car body shells. Recently, carbon fibres have been used instead of glass in some instances to produce a stronger, more lightweight product. Carbon fibre is used extensively in Formula One racing cars.

Concrete

Concrete (a material produced by mixing together sand, cement, gravel and water) is good in compression, but very weak in tension. To produce a better all-round product, concrete is cast around a cage of pre-stressed, steel reinforcements. The steel gives added strength by taking up the tension stresses, while the concrete takes up the compression stresses. In this way, both materials make a contribution to the performance of the composite material.

C *The Colorado Street Bridge, a historic concrete bridge in Pasadena, USA*

Textile-based materials

Many modern textiles are made of combinations of materials and can be 'designed' for specific tasks. Many manufacturers blend fibres to increase their functionality and diversify their use. Blending consists of two or more different types of fibres twisted or spun together, and can include natural, regenerated and synthetic fibres. Blending produces a **combined material** where the properties of each contributing material are enhanced.

Common fabric blends

There are several common fabric blends.

Polyester and cotton: the crease-resistance of polyester combines with the comfort of cotton. It is easily laundered, dries quickly and is ironed at a lower temperature than pure cotton.

Nylon and wool: the blending of nylon with wool makes this fabric more absorbent and softer. It becomes stronger and more durable.

Nylon and acetate: this combination makes the fabric more absorbent than nylon alone.

Silk and wool: the blending of silk with wool gives a subtle texture to the fabric.

Cotton and polyester: polyester gives cotton a permanent crease. The fabric is extremely soft, resists wrinkling and is easy to care for. It is used widely for shirts.

Coated fabrics

Coated fabrics are flexible combined materials consisting of a woven or non-woven textile with a coating of plastic or other material, designed for specific applications. Coated fabrics are used for products such as hot air balloons, airbags, luggage and clothing.

Objectives

Understand that materials can be combined and processed to create more useful, or desirable, materials or products.

Understand that many textile fabrics are blends of different fibres that can be adjusted to create different properties for different purposes.

Understand that when using food, combining and processing materials can provide outcomes with different working characteristics.

Key terms

Combined material: A material made from two of more different materials, with each material still retaining its own identity in the new material.

Working characteristics: The way a material behaves when it is shaped and formed.

∞links

To find out more about textile fibres, see 10.8.

To find out more about carbon fibre and Kevlar, see 11.5.

A *Examples of products using coated fabrics*

Technical textiles

Technical textiles are manufactured mainly for their performance and functional properties, and can include other materials. Technical textiles may not always be visible in a product, but they can improve the products performance.

- Carbon fibre, a synthetic material, is a major component in aircraft wings.
- Kevlar is used to manufacturer protective body armour.
- The use of advanced fabrics in car tyres contributes to performance, road handling and tyre durability.

Food products

When designing food products it is important to understand how individual ingredients will react when combined or processed. A clear understanding of the **working characteristics** of ingredients will help the food designer develop a successful product.

The following are methods by which the properties of foods can be changed or modified.

Aeration: a way of making foods lighter, which improves its texture. Margarine and egg whites are used for this.

Thickening and setting: a technique used to thicken sauces and 'set' products such as quiches. It uses eggs and starches.

Shortening: a means of making food 'crumbly' by using fat.

Binding: a way to make foods stick together. Eggs, milk or water can be used.

Colour: the appearance of foods can be improved by using sugar, fruit or vegetable.

Flavour: the taste of foods can be modified with herbs, spices, sugar, fruit or vegetables.

Emulsifying: a way of stopping fat or oil from separating from a mixture. Egg yolk is used for this purpose.

B *Product designed around technical textiles*

∞links

To find out more about technical textiles, see

www.futuretextiles.co.uk/index.cfm

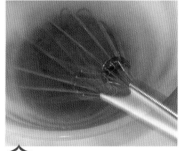

C *Chefs combine and process ingredients*

Summary

Blending textiles produces a combined material where the properties of each contributing material are enhanced.

Fabrics can be coated to produce combined materials for specific purposes.

Technical textiles have many high performance applications.

Ingredients used in the manufacture of food products can have their working characteristics modified.

Activities

1. 'Blending' is often used in the textiles industry to produce a fabric made up of two or more fibres twisted or spun together. For each of the following blends, describe its characteristics, and how these characteristics differ from the unblended fibre:
   a polyester and cotton   b nylon and wool   c linen and silk.

2. Explain the effect of mixing elastane with cotton when weaving a fabric.

3. Describe a food product you have made by combining ingredients. List the ingredients, and explain what each ingredient contributes to the final product.

11.4 Smart materials and smart systems

Product designers have a wide range of materials available to them. But imagine the possibilities of using materials which have **properties** that change. Materials that respond to changes in their environment are called smart materials. Smart systems are products designed to respond to changes in their environment, but they are made from more than one material.

Smart materials and smart systems allow the product designer to develop products with greater levels of **functionality**. They also allow existing products to be improved. Using smart materials instead of conventional mechanisms can simplify devices, reducing weight and the chance of failure.

Smart materials

Smart materials can be grouped by how they react to their environment.

Colour-changing materials

- Photochromic materials change colour in response to changes in light. Some glasses use reactive lenses, which become darker as the light levels increase.

A *Glasses with reactive lenses*

- Thermochromic materials change colour in response to changes in temperature. This technology has been used to produce bath plugs that change colour when the water is too hot, and clothing that reacts to the body temperature of the wearer.

Light-emitting materials

- Electroluminescent materials can produce brilliant light of different colours when an electric current is passed through them. Light generated in this way produces no heat, and has been used to illuminate emergency exits in public buildings.
- Fluorescent materials produce light when exposed to UV-rays. The light stops when the UV radiation is removed.
- Phosphorescent materials produce light as a result of being exposed to a light source, but only after the source has been removed. Emergency warning signs that can be read 24 hours a day have been developed using this technology.

Moving materials

- Piezoelectric materials transform mechanical energy to electrical energy and vice versa. This technology has been used to produce 'smart skis'.

B *Head i XRC 300 Ski with intelligent technology*

Objectives

Understand that the development of smart materials allows the designer to meet a variety of user needs in new and exciting ways.

Have an insight into the range of smart materials available to a designer in the 21st century, and their commercial impact.

Key terms

Properties: how materials perform in everyday use.

Functionality: what a product does.

UV (ultraviolet) rays: a light source with a shorter wavelength than visible light.

⚭ links

To find out more about the mechanical and physical properties of materials, see 10.1 and 10.2.

To see how a textile designer has used smart technology, take a look at www.sablechaud.eu

- Shape-memory alloys are metals that, after being strained, revert back to their original shape at a certain temperature. Applications include shape memory stents (tubes threaded into arteries that expand on heating to body temperature to allow increased blood flow) and Memoflex glasses.

- Quantum tunnelling composites have been used to create touch-sensitive switches to control electronic devices in clothing.

Temperature-changing materials

- Thermoelectric materials act like a heat pump. When a current is applied, one side of the material cools down while the other heats up. The effect is reversible. A plate has been developed using this technology that can keep foods either cold or warm depending on the polarity of the power source.

Thickness-changing materials

- Magneto-rheological fluids become solid when placed in a magnetic field. They can be used to construct dampers that suppress vibrations. These dampers are often fitted to buildings and bridges to suppress the damaging effects of high winds or earthquakes.

The future of smart materials and smart systems

The development of smart materials and systems will create an exciting new generation of products.

- With an ageing population, there is a growing market for products that make life easier for the elderly. Using shape memory materials, food packaging is being developed that automatically opens on heating for people with arthritis.

- Research is currently being carried out into the production of electronic products that would incorporate a technology known as 'active-disassembly'. In this process, fasteners would be constructed from shape memory materials. When the time comes for the product to be recycled, the product would be heated and the individual components would 'self-release'. The components could then be separated by simply shaking the product, saving time and money. By using fasteners that react to different temperatures, products could be sorted in a pre-determined pattern. This technology could be in use in the next two years for products such as mobile telephones.

Nanotechnology

Nanotechnology, or 'nanotech' involves working with materials on a very small scale, less than 100 nanometres in size (the average human hair is about 80,000 nanometres wide). Nanotechnology can be described as the manipulation of individual atoms to modify a material's properties. This is called molecular manufacturing. Nanotechnology is used in many commercial products and processes. It is difficult to predict how this technology will develop, but scientists have suggested that we could, in the future, see the development of incredibly tiny machines that are too small to see: the ultimate smart systems!

∞links

To find out more about smart skis, and about quantum tunnelling composites, see 11.6.

Activities

1 What is meant by the term 'smart' material?

2 Name, and describe, a product you are familiar with which uses a 'smart' material.

3 Choose a product that you are familiar with and suggest ways in which it could be improved with smart materials.

4 There is a growing demand for 'intelligent fabrics'. What do you think this term means?

Study tip

Can you explain what smart materials are and some of their uses in new products? It pays to have an awareness of as many as possible.

Summary

Smart materials react to changes in their environment.

Smart systems are products designed to react to changes in their environment.

Smart materials and smart systems allow products to be designed with greater functionality.

Modern materials

Modern materials are considered to be those that have been developed over the past 50 years. These materials are different to smart materials because they do not react, or respond, to their environment.

Some modern materials can trace their origins back to traditional materials, while others are completely new. Many consist of a mixture of materials, and can be classed as **composite materials**.

Modern materials in common use

Carbon fibre

Carbon fibre consists of extremely thin fibres of carbon, twisted together to form a yarn, which is then woven to create into a fabric cloth. Carbon fibre cloth can be combined with a plastic resin, which bonds the fibres together, to create a very strong composite material for use in applications where a high strength, low weight material is required. Carbon fibre is used for Formula One racing cars, technical cycles and helicopter rotor blades.

A *A carbon fibre cycle frame*

Fibre optics

Fibre optics is the science of using glass fibre or plastic fibre to transmit light. Fibre optic cables enable data to be transmitted over large distances more efficiently than other forms of communication. One glass fibre could replace hundreds of copper cables. Fibre optic cables are also widely used in lighting applications.

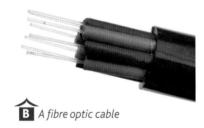

B *A fibre optic cable*

Waterproof and breathable

Waterproof of breathable fabrics (such as Gore-tex and Sympatex) consist of a porous membrane, laminated to high-performance textiles. This membrane contains billions of microscopic pores which allow moisture vapour to escape, but which water droplets cannot penetrate. Laminating produces a more versatile fabric in which the outer laminates protect the easily damaged membrane. This material is widely used to produce clothing for activities such as walking and cycling.

C *Waterproof and breathable clothing*

Objectives

Understand that the development of modern materials has allowed the designer to meet a variety of user needs in new and exciting ways.

Develop an awareness of how modern materials are making an impact in commercial products.

Key term

Composite materials: materials made from two more different materials, with each material still retaining its own identity in the new material.

∞ links

For more information on smart materials, refer back to 11.4.

Kevlar

This has a wide range of applications, from bicycle tyres to body armour such as stab vests. Weight for weight, it is five times stronger than steel. Kevlar is a flexible, plastic based-material, and was developed in 1965. It is usually woven and used in layers.

Maplex

This is a plain brown board, similar to plywood. It is a completely sustainable material, made of 100 per cent pressed wood fibres using a process that requires no binders or glues. Instead, only water, heat and pressure are used. Maplex is strong, easily moulded and completely biodegradable.

Polycaprolactone (PCL)

PCL is a low melting point, biodegradable thermoplastic material, widely used in medical applications. Known in schools as polymorph, it can easily be shaped by hand to resemble injection-moulded products.

Precious metal clay (PMC)

PMC is a material mainly used by jewellers, and was developed in the 1990s. Microscopic particles of silver or gold are combined with a moist binder to create a material that has the feel and working properties of modelling clay. Designs are shaped and moulded in a similar way to traditional clay, air-dried to remove any moisture and then fired to allow the metal particles to fuse together to make a dense, fully metallic object. After it has been removed from the kiln and allowed to cool, finished PMC jewellery can be treated like any other silver or gold item.

D A Kevlar helmet

E A Maplex chair

F Polymorph

G Silver PMC jewellery

Study tip

Modern materials are similar to smart materials but do you know the difference?

Activities

1   Hill walkers often use waterproof garments that contain Gore-tex. What does this modern material do, and how does it work?

2   Carbon fibre is often used in applications such as Formula One racing. Explain what carbon fibre is, and how it contributes to this particular sport.

Summary

Modern materials are those that have been developed over the past 50 years.

Many modern materials are a mixture of materials, and are often classed as composite materials.

kerboodle

Smart systems on the slopes

Objectives

Be aware of how product designers are using smart and modern materials to shape and develop products for the 21st century.

Case study

Smart skis

When skiing at high speeds and on tough terrain, skis tend to vibrate, lessening the contact area between the ski edge and the snow surface. This results in reduced stability and control, and decreases the skier's speed. Smart skis incorporate vibration control technology, and allow one pair of skis to perform almost perfectly on all types of snow.

Each smart ski has a microchip-based control system and ceramic **piezoelectric** sensors built in, to monitor how well the ski is performing. When the piezoelectric sensors detect vibration, they send an electrical signal to the control circuit, which in turn sends pulses of electric energy into the piezoelectric material to change its stiffness and dampen the unwanted vibrations. The piezoelectric material act as switches, quickly cancelling out vibrations by either relaxing or stiffening the ski, giving the skier more control and a smoother ride. This process takes less than 5 milliseconds.

B *Smart skis in action*

non-polarized piezoelectric material

polarized piezoelectric material

piezoelectric Ski remains stable

forces vibrate ordinary ski

A *How the sensors work*

Case study

D30 protective beanie

Not everyone who skis or snowboards wants to wear a helmet. But with the introduction of D30, it is now possible to wear head protection that is a little more stylish. D30 is a specially engineered material made with intelligent molecules. The beanie shown on the right has a D30 insert that is soft and flexible, but on impact the molecules lock together to absorb shock in a similar way to rigid helmets. D30 is used in a wide range of products.

C *A protective beanie showing the D30 insert*

⊙⊙**links**

To find out more about D30 watch a video at: **www.d3o.com**

⊙⊙**links**

To find out more about softswitch technology, click on QTC technology at **www.peratech.com**

▮ Modern materials get the boot

Nike SL football boots

Traditionally, football boots were made from leather. Developments using synthetic materials, mainly plastics, enabled interesting shapes to be produced. In 1996, Adidas produced the Predator, which used rubber ridges to help swerve the ball. Recently, Nike was asked to produce the fastest and lightest boot possible. After three years of research, the Nike SL was launched, which weighs only 190 grams and is made almost entirely of carbon fibre.

D *Nike SL carbon fibre football boots*

▮ Put your finger on the button

Softswitch wearable technology

Quantum tunnelling composites (QTCs) have the ability to change from being an insulator to a conductor when pressure is applied. QTCs contain particles of metal held in a non-conducting thermoplastic material. When the material is squeezed, the metal particles move closer together and the material becomes a conductor. QTC controllers have been integrated into clothing to make it more convenient to operate devices remotely.

iPod

Bluetooth chip

Data Ribbon

Keypad

E *Wearable technology designed around a softswitch*

Key terms

Piezoelectric: describes the ability of some materials to generate electricity when deformed by mechanical pressure. This effect is also reversible, causing piezoelectric materials to deform when a small voltage is applied.

Quantum tunnelling composites (QTCs): smart materials with unusual electrical properties. QTCs, in their normal state, act as insulators, but when compressed they conduct electricity.

Study tip

Use the Case studies to help you to remember some smart and modern materials and their uses in new products. Can you write your own case study about a product you like which uses smart or modern materials?

Activity

Consider the technologies described on the previous pages, and design a future product which could take advantage of these developments.

Summary

Four Case studies show how smart and modern materials have been incorporated into new products.

11.7 Standard components

Standard components are pre-manufactured parts, bought in by individuals or commercial manufacturers, usually as part of a batch production or mass production system.

Whenever you solder a resistor to a circuit board, sew a piece of Velcro onto a garment or use a jar of curry sauce you are using a standard component. On a larger scale, standard components are an essential part of modern manufacturing. Many of the items we use everyday contain these pre-manufactured parts.

The standardisation of components brings many benefits to the producer and the consumer, including a reduction in manufacturing costs; standard sizing; consistent quality; more efficient production; reduced maintenance costs; reduced waste and more scope for recycling.

To maximise these benefits, the product designer needs to consider the use of standard components at the design stage of a product to help plan the manufacturing process.

Standard components are essential in a global market. **Economies of scale** would not be possible if products were limited to their home markets. A product that could only be used in its country of manufacture would have a limited market. Today, it is possible to buy products that are used all over the world, and this has led to a greater use of standard components. For example, most MP3 players use a standard USB socket to connect to a computer to download music. The USB socket would be made in a factory dedicated to making this component and then bought-in by different MP3 manufacturers.

This approach is widely used in the car industry, where one manufacturer might use a standard component in a range of vehicles. The Volkswagen Group includes Volkswagen, Audi, Seat and Skoda and, although the products of these various companies appear different, many of them share the same parts and are produced by specialist suppliers.

Objectives

Understand how pre-manufactured standard components are used to improve the effectiveness of the manufacturing process.

Be able to identify a range of components appropriate to the material areas you have studied.

Key terms

Economies of scale: mass production allows products to be produced at a lower cost.

⚭ links

For more about sustainability and product life cycles, see 6.1 and 6.2.

For information on scales of production, take a look at 7.1 and 7.2.

A *Different brands of car can share many of the same parts*

The main advantages of using standard components are that they are quick and easy to use in many different products in a variety of ways. They can be mass-produced so that they can be made cheaply in large quantities.

Many electronic products rely on a wide range of standard components, including LEDs, transistors and integrated circuits. Textile products make use of standard components, such as Velcro, buttons, zips and press-studs. Nails, screws, bolts, rivets and hinges are just some of the components used in manufacturing. Standard components are widely used in the food industry as they help to ensure that products are consistent. Examples of standard components used in food production include: stock cubes, marzipan, icing, pasta sauces, frozen garden peas, pre-cooked prawns, pizza bases and readymade pastry cases for flans and quiches.

Activity

What are the benefits to the manufacturer of buying-in pre-manufactured components? Illustrate your answer by referring to a product of your choice, and naming the standard component used.

B *Standard components*

Study tip

Standard components are used in many products. What standard components have you used in your project work at school? Write about your own experiences of designing and making products at school.

Summary

Standard components are pre-manufactured parts used by individuals (for example, buttons on a hand-knitted cardigan) or large manufacturers (for example, USB sockets on MP3 players).

Standard components play an important part in many manufacturing industries, including the electronics, food and textiles industries.

Practice questions

1   Materials are classified into different groups according to their properties.
  (a) (i) Name a material which is renewable. *(1 mark)*
      (ii) Name a material which is non-renewable. *(1 mark)*
      (iii) Explain the difference between renewable and non-renewable materials. *(2 marks)*

  (b) Circle the correct classification for four of the following materials: *(4 marks)*

Materials	Classification		
Copper	Non-ferrous metal	Ferrous metal	Alloy
High impact polystyrene	Thermosetting plastic	Natural plastic	Thermoplastic
Slip	Body clay	Liquid clay	Dried clay
Silk	Regenerated Fibre	Natural fibre	Synthetic fibre
Plywood	Man-made board	Hardwood	Softwood
Polyester	Regenerated Fibre	Natural fibre	Synthetic fibre
Rice	Vegetable	Protein	Cereal
Foil lined Duplex board	Composite	Paper	Cardboard

  (c) Explain how trees are turned into paper/card. *(3 marks)*

  (d) Explain why standard components are important when manufacturing products. *(2 marks)*

2   Materials are available in a variety of shapes, forms and sizes. These are often known as stock forms.
  (a) Name a material and give two examples of stock forms it can be purchased in.
      (i)   Material … *(1 mark)*
      (ii) Stock form 1… *(1 mark)*
      (iii) Stock form 2 … *(1 mark)*

  (b) Explain why materials come in stock forms. *(3 marks)*

  (c) Raw materials are turned into useful stock sizes through primary processing. One example of this is turning durum wheat into pasta.

      Using notes and sketches, explain the primary processing of a material you have studied. *(6 marks)*

3   Materials are often combined when manufacturing products.
  (a) The table below shows different types of material. Choose two types of material and complete the row. The first material in each row is completed for you. *(6 marks)*

Type of material	Material name 1 (example)	Material name 2	Material name 3	Material name 4
Paper/card	Duplex board			
Timber	Oak			
Metal	Copper			
Plastic	Acrylic			
Ceramics	Stoneware			
Food	Cereal			
Textiles	Wool			

(i)   Name a material which has been created from two or more other materials.   *(1 mark)*

(ii) What is the word used to classify this material?   *(1 mark)*

(iii) Explain why materials are often combined in this way.   *(3 marks)*

4   Select two of the products below.

List two properties of the material that they are made from and state whether or not they are suitable for recycling.

Cotton shirt

Cheese

Porcelain teaset

Laminated playing cards

Aluminium alloy scooter

Polypropylene bath toy

Plywood skateboard

(i)   Name of product ...
List two properties of the material from which it is made.   *(2 marks)*
Is the material suitable for recycling?   *(1 mark)*

(ii) Name of product ...
List two properties of the material from which it is made.   *(2 marks)*
Is the material suitable for recycling?   *(1 mark)*

(b) New materials are developed using new technologies and enable manufacturers to design innovative products.

(i)   Name a new material you have studied.   *(1 mark)*

(ii) Describe its properties and using specific examples, explain how these properties are used to design and manufacture new products.   *(6 marks)*

Objectives

By reading these pages you should gain a basic understanding of what your Controlled Assessment project needs to be like.

■ What is this unit all about?

You will be tested on your ability to design and make a prototype for a new product through a process now called Controlled Assessment. In the past, this would have been known as coursework, but Controlled Assessment is slightly different.

The task you undertake must be chosen from a list published by AQA. These tasks will allow you a fair amount of freedom but have been written to ensure that you can access all of the **assessment criteria** and are really suitable for this course.

A consequence of Controlled Assessment is that the work will be closely supervised by your teacher with most of the work being carried out in school. This will mean that there will be restrictions upon what work you are allowed to complete at home and your homework tasks are likely to be related more to planning what you will need to be effective in the next D&T lesson or related to the written paper. It will, of course, mean that for every lesson you will need to ensure that you are making full use of the time available.

You will need to produce a design folder or journal that shows all your investigations and decision-making. This can be presented in a wide range of different styles, including A3 or A4 folders, sketchbooks or electronic portfolios. The information needs to be concise and clear for somebody else to follow. You will be rewarded with additional marks if it is presented in a clear and concise manner. As a guide, the design folder should be around 20 sheets of A3 paper, with drawings, notes and photographs – or the equivalent, if presented in a different format.

In addition to the design folder, you will need to manufacture a prototype that is well made and as close as you can make it to how it would appear in a shop. Some students consider the packaging and/or display material needed to launch this new product. You will need to demonstrate your very best making skills by either showing that you are very skilled in a narrow range of processes or that you have a broad range of skills using a wider range of processes. You must provide good photographic evidence of your making. This should cover all the stages in your work, including photographs of the final product. Make sure these photographs show close-up details of your work.

It is expected that you spend approximately 45 hours on this Controlled Assessment. Your teacher will explain in more detail the schedule for completing the work, which will vary between schools.

Key terms

Assessment criteria: this is a list used by your teacher as a guide when marking your work.

Moderator: this person will be a teacher from another school who has been trained by AQA to ensure that the marks your teacher awards are in line with AQA standards.

The five areas you are being assessed on are worth 90 marks in total. These are called assessment criteria and are broken down as following:

- Investigating the design context   *8 marks*
- Development of design proposals   *32 marks*
- Making   *32 marks*
- Testing and evaluation   *12 marks*
- Communication   *6 marks*

Your teacher will mark your work initially. However, the moderator will request a sample of your work and your marks may go up or down as a result of this check. It is very important to remember that the moderator will not have an opportunity to talk to you, so you must provide as much evidence of the work you have done as you can. It is vital that you do not throw anything away, such as rough drawings or models, and that you present your work in a way that is easy to follow.

This unit will help you to plan your work and compare it to good examples of work done by students in other schools.

Activity

Look at an example of work from a past student (you could use one of the examples provided by AQA or one provided by your teacher) and mark this against the assessment criteria. Make a list of what would be needed to achieve the top grade.

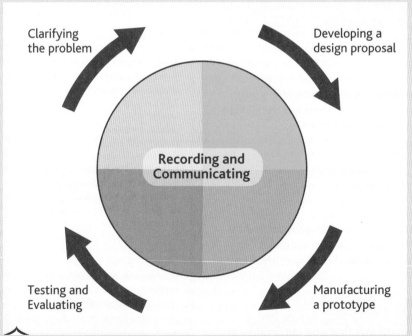

A *Designing and making is not a linear process, so the order of work will vary, but you will need to provide evidence of work in each of these areas*

12.1 Investigating the design opportunity

■ What is a design opportunity?

A design opportunity might be thought of as the situation that causes a problem, which can then be solved by making a new product. It might also be thought of as an opportunity to create a product that is totally new, or at least different from what already exists. AQA will provide a range of suitable design opportunities. Your first job will be to turn one of these into a specific design brief that you can work to. In most cases that will mean undertaking some initial research in order to narrow the task down into something you can complete within the time available.

The evidence you will need to show for this part of the project will include:

- How you plan, collect and select the relevant research material.
- How you explain your understanding about the design opportunity you have chosen.
- An investigation into similar products or ones that will help you to better understand what can be manufactured.
- Your analysis of the research material you have collected.
- A clear design brief, which fully explains who the target user would be, where your product might be sold and what quantity would need to be manufactured commercially.

Writing a design brief

You must assume that the product you design and prototype is going to be commercially manufactured. You should try to explain the design opportunity simply and might start with the basic sentence:

'I intend to design and make ... .'

You should try to explain the sort of person who would use your product. What age range would they fit into? What sort of lifestyle would they live?

You should try to explain the sort of environment your product would be used in and any issues that this would highlight. For example, children's toys will need to stand up to a great deal of abuse; they will need to be virtually indestructible!

You will need to consider a wide range of issues, such as costs, packaging, material properties, environmental issues, safety, and a whole host of other factors, but if there are any that you can highlight at this stage it is useful. For example, you might have a target price range or the product might need to make use of recycled materials.

Objectives

Be able to describe fully, at the start of your design folder, exactly what you hope to achieve in your project.

links

Remind yourself of some of the issues around designing new products. See 2.1 for information on technology push and market pull, and 2.3 for more about continuous improvement.

Chapter 6 (6.1 and 6.2) give you information about global responsibility and sustainability, which are becoming increasingly important considerations.

Creative play in the under fives is an important part of their development. A leading retailer has asked me to investigate the market for low priced dressing-up outfits which will encourage children to take on a variety of roles. I have been asked to design and prototype a complete oufit and the first batch of 1000 outfits will be produced in three different sizes. I will need to consider packaging which will need to reflect the corporate identity of the retailer. I will also provide an information leaflet to parents on how to encourage creative play through using the outfit.

A A clear design brief which doesn't suggest a solution at this stage

Identifying the target user

Many students are able to narrow down their **target market** at a very early stage and present their thoughts in a visual way. This is sometimes known as a **client profile** and can be useful throughout the designing to act as a reminder of who would use the product, and possibly whether the style of the product would be suitable.

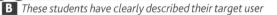

B *These students have clearly described their target user*

Commentary

Ensure that you can provide information on the target user at an early stage of the design folder as this will help the moderator to assess whether your design ideas are suitable.

12.2　Planning your research

It is important that you are able to plan what tasks you have got to do and what sort of research is likely to be most suitable. Jotting down what you already know about this design opportunity is a good starting point. Making a list of what you need to find out is a vital step forward. It is always best to seek out sources for **primary research** whenever that is possible. It is important to remember that researching should allow you to move forward quickly to designing.

The importance of research is represented in Diagram **A**.

Organising how you will gather and order your research findings is an important task. Remember that you are likely to have to undertake some additional research at the development stage of your project. For example, that would be an appropriate time to look into the suitability of specific materials and manufacturing processes.

◼ What sort of research is expected?

This very much depends on the sort of project you are undertaking and your teacher will be able to advise you on what might be most relevant to you. But as a general guide, the following areas might be considered:

- similar products which already exist on the market (see the section on product analysis on page 140)
- what the target market would be most interested in
- visual resources that might help you develop the shapes and style of the product
- anthropometric/ergonomic information relevant to the product
- manufacturing processes that you might need to try out to see whether your initial ideas are feasible
- materials – do not attempt to do this until the development stage when you know the properties needed for the product
- construction and finishes – again, do not attempt to do this before the development stages.

The four Ps

These can be used to help organise your research.

People

Who do you need to talk to? These might include experts such as a nursery teacher, parent and shopkeeper or people from the target market.

Places

Where do you need to visit? This might include shops, museums, art galleries, libraries and the internet or might be related to the 'People' section. Visiting a nursery to observe children playing would be a sensible starting point for designing play equipment, for example.

Objectives

Understand why research and planning are important before you start actually designing.

Understand how to show your teacher what preparation you have been doing.

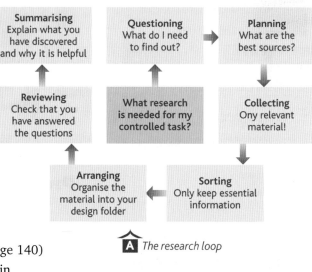

A *The research loop*

⬭⬭ links

For more information on anthropometrics and other human factors, see 2.2.

For some tips on presentation techniques, see 16.2.

Commentary

Do not copy general information on materials and constructions.

Make sure that you analyse – not simply describe – what you have found out.

Always explain why the information is useful to you.

Products

Undertaking product analysis is probably the best form of research. Analysis is more than describing. Think of it as a discussion. You will need to examine different viewpoints. Looking at the work of other designers (**secondary research**) is also a very useful starting point. Looking at other unrelated products or buildings might provide some really useful visual or constructional information.

Processes

Almost all useful research under this heading would be research such as practical experiments and tests. There is little value in copying information out of books or from the internet.

Key terms

Primary research: this is research where you have found out the information yourself. This might include, for example, interviews with experts, material tests, observations and product analysis.

Secondary research: This is where you use material that someone else has put together (for example, copying information on anthropometrics, materials and manufacturing techniques).

Presenting your research

You may need to collect information from a wide range of sources and much of this will need to be done outside of lessons. You should keep a list of all the research you undertake but you do not necessarily need to present all of this material. You could present the information about the range of research you have undertaken in the form of a chart.

It is likely that your teacher will provide you with deadlines for completing specific research. Make sure that you have collected and sorted the material before the lesson and have made notes about why it is useful to you. If you cannot do that then it is likely that you will gain little credit for what you have collected.

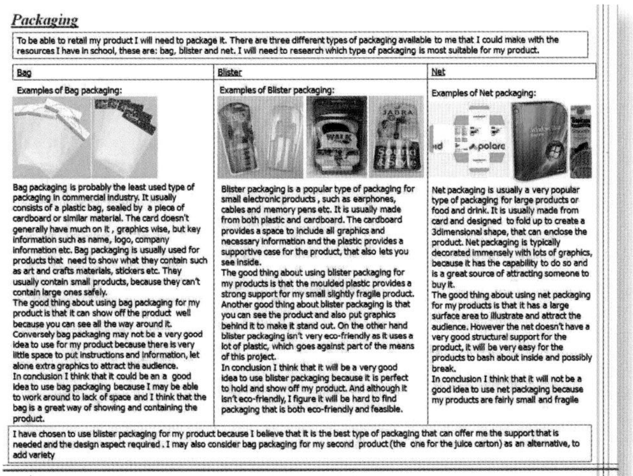

Packaging

To be able to retail my product I will need to package it. There are three different types of packaging available to me that I could make with the resources I have in school, these are: bag, blister and net. I will need to research which type of packaging is most suitable for my product.

Bag

Examples of Bag packaging:

Bag packaging is probably the least used type of packaging in commercial industry. It usually consists of a plastic bag, sealed by a piece of cardboard or similar material. The card doesn't generally have much on it , graphics wise, but key information such as name, logo, company information etc. Bag packaging is usually used for products that need to show what they contain such as art and crafts materials, stickers etc. They usually contain small products, because they can't contain large ones safely.
The good thing about using bag packaging for my product is that it can show off the product well because you can see all the way around it.
Conversely bag packaging may not be a very good idea to use for my product because there is very little space to put instructions and information, let alone extra graphics to attract the audience.
In conclusion I think that it could be an a good idea to use bag packaging because I may be able to work around to lack of space and I think that the bag is a great way of showing and containing the product.

Blister

Examples of Blister packaging:

Blister packaging is a popular type of packaging for small electronic products , such as earphones, cables and memory pens etc. It is usually made from both plastic and cardboard. The cardboard provides a space to include all graphics and necessary information and the plastic provides a supportive case for the product, that also lets you see inside.
The good thing about using blister packaging for my products is that the moulded plastic provides a strong support for my small slightly fragile product. Another good thing about blister packaging is that you can see the product and also put graphics behind it to make it stand out. On the other hand blister packaging isn't very eco-friendly as it uses a lot of plastic, which goes against part of the means of this project.
In conclusion I think that it will be a very good idea to use blister packaging because it is perfect to hold and show off my product. And although it isn't eco-friendly, I figure it will be hard to find packaging that is both eco-friendly and feasible.

Net

Examples of Net packaging:

Net packaging is usually a very popular type of packaging for large products or food and drink. It is usually made from card and designed to fold up to create a 3dimensional shape, that can enclose the product. Net packaging is typically decorated immensely with lots of graphics, because it has the capability to do so and is a great source of attracting someone to buy it.
The good thing about using net packaging for my products is that it has a large surface area to illustrate and attract the audience. However the net doesn't have a very good structural support for the product, it will be very easy for the products to bash about inside and possibly break.
In conclusion I think that it will not be a good idea to use net packaging because my products are fairly small and fragile

I have chosen to use blister packaging for my product because I believe that it is the best type of packaging that can offer me the support that is needed and the design aspect required . I may also consider bag packaging for my second product (the one for the juice carton) as an alternative, to add variety

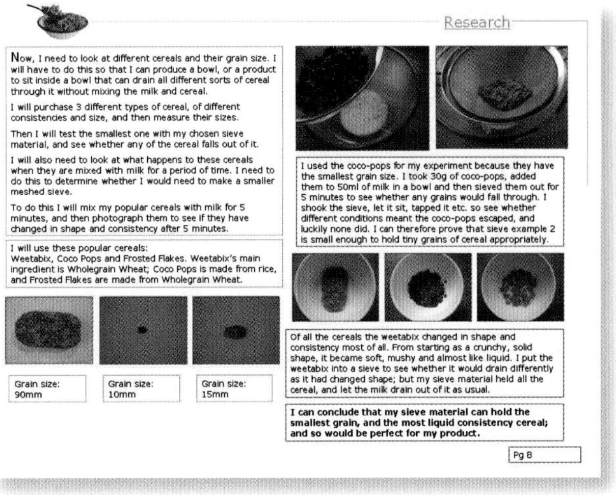

Research

Now, I need to look at different cereals and their grain size. I will have to do this so that I can produce a bowl, or a product to sit inside a bowl that can drain all different sorts of cereal through it without mixing the milk and cereal.

I will purchase 3 different types of cereal, of different consistencies and size, and then measure their sizes.

Then I will test the smallest one with my chosen sieve material, and see whether any of the cereal falls out of it.

I will also need to look at what happens to these cereals when they are mixed with milk for a period of time. I need to do this to determine whether I would need to make a smaller meshed sieve.

To do this I will mix my popular cereals with milk for 5 minutes, and then photograph them to see if they have changed in shape and consistency after 5 minutes.

I will use these popular cereals:
Weetabix, Coco Pops and Frosted Flakes. Weetabix's main ingredient is Wholegrain Wheat; Coco Pops is made from rice, and Frosted Flakes are made from Wholegrain Wheat.

Grain size: 90mm

Grain size: 10mm

Grain size: 15mm

I used the coco-pops for my experiment because they have the smallest grain size. I took 30g of coco-pops, added them to 50ml of milk in a bowl and then sieved them out for 5 minutes to see whether any grains would fall through. I shook the sieve, let it sit, tapped it etc. so see whether different conditions meant the coco-pops escaped, and luckily none did. I can therefore prove that sieve example 2 is small enough to hold tiny grains of cereal appropriately.

Of all the cereals the weetabix changed in shape and consistency most of all. From starting as a crunchy, solid shape, it became soft, mushy and almost like liquid. I put the weetabix into a sieve to see whether it would drain differently as it had changed shape; but my sieve material held all the cereal, and let the milk drain out of it as usual.

I can conclude that my sieve material can hold the smallest grain, and the most liquid consistency cereal; and so would be perfect for my product.

Pg 8

B *These students have recorded the results of their research, explaining what they found out and what was particularly useful to the project*

Activity

Keep a notebook of the processes you have used in the past – both during KS3 and the early part of the GCSE course. Include photos of the equipment and describe the process in detail. This is primary research and could be useful when presenting your research findings.

Learning outcomes

After reading these pages, you should have an understanding of why it is important to spend time planning and doing initial research before you start designing.

You will also be able to share information with your teacher about the preparatory work you have been doing.

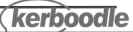

Making a detailed examination of existing, commercially made products can provide you with some of the most useful information to aid you when designing. Sometimes it is useful to look at similar commercially made products so that you can fully understand the reasons the products are like they are. Sometimes it is more useful to look at completely different products as a way of generating new ideas.

When looking at similar products, think about the **design and manufacturing considerations** listed below:

Materials
What are the properties of the materials the product is made from and why are these properties important?

Function
What is the product designed to achieve? How does it work? Are there any scientific principles involved?

Product maintenance
How can it be maintained? Are there issues such as battery access that needed to be considered? Have the manufacturers used standard components and why was this important?

Target market
Who is it designed for? Where would it be sold? How would the consumer get the product home? Is it ready to use or does it need to be assembled at home? Are there instructions provided and are they easy to follow?

Ergonomics
Has the product been well designed to suit the user you have identified as the target user? If not, then what changes would be needed to make it more suitable? Are there any safety considerations that needed to be addressed?

Manufacturing
How was it commercially manufactured? How can you produce something similar in school?

Style
How can you describe the style? For example, is it modern or traditional? What words could you use to describe how you feel about the product? For example, is it feminine, masculine, aggressive, utilitarian, trendy, futuristic? Our emotional responses to products are very important.

Packaging
How was it packaged and transported to the retailer? What specific jobs does the packaging do? Can this be reduced?

Environmental impact
Are there any issues that will have a serious impact on the environment? This could be the types of materials used, and the amount of energy that has been consumed in the processing and transportation of the product. You might consider whether the product has been designed to last a long time or whether it has been made with designed obsolescence in mind.

Costs
What is the selling price of the product? Can you suggest reasons for this and whether costs could be reduced by altering the design? Can you explain whether the level of quality matches the price?

Objectives

Be able to undertake your own product analysis in enough depth for you to move your project forward quickly.

Use your own product analysis when you write design criteria for your own product.

Key terms

Design considerations: areas that need to be considered when designing. These are often decided from analysing existing products and would largely be concerned with the user and areas, such as aesthetics, safety, usability and costs.

Manufacturing considerations: areas that need to be considered when manufacturing. These are often decided by analysing existing products and would depend largely upon the scale of production required and the material properties.

Aesthetics: the features in a product that make it visually appealing, such as colour, texture, shaping and styling features.

∞ links

Look at 1.1, 2.3 and 3.1 to see how product analysis of different products can help when generating new ideas.

Product Analysis

Target Market
Aimed at young children, can be used as a night light so they are not scared of the dark

Ergonomics
There is a Cord Switch that makes it easy to switch the lamp on off. The bulb is not easy to change as you have to take the two layers of plastic off to get access to the bulb.

Packaging
Minimal triangular shaped packaging that is designed for the lamp so that the lamp does not move around and is compact inside the packaging.

This is the packaging which will carry the lamp which will be dissembled.

The foot is shaped in this way so that the lamp is stable, this also makes the design look more futuristic.

These are the fins at the top of the lamp so that air can come through whilst it is spinning.

This is where the bulb is placed and the plastic cover sits on top of this.

Materials
This Blimp child's lamp from IKEA. The lamp shade is made up of two layers. The first layer is made from a thin lightweight layer of polypropylene plastic. The outer diffusor is made from PET plastic. The feet are also made from polypropylene. The legs and wire frame are made from steel which have been powder coated.

Aesthetics
The product has a very simple and effective design. But I think that the products main strength is the idea of the heat rising from the bulb, which then rotates the central polypropylene cylinder. The design is the reflected on to the outer skin.

Safety
There are no sharp ends and it is safe for young children, who this product is aimed at. It is safe turn the lamp on and off from the Cord Switch. The shade does not get too hot as there are slits in the top for the heat to escape.

Size
Diameter: 14cm
Height: 19cm
Cord Length: 2.5cm

Materials
This lamp is made from wood and metal, which make up the structure of the lamp. It has two shades. The inner made from polypropylene that has had a design printed on to it. The outer diffusor is made from a laminated piece of plastic and a thin fabric like material.

Aesthetics
The lamp is cuboid in shape. The design changes on each face as the inner cylindrical layer rotates. The lamp is an interesting for children because it is colourful and it is based on sea life that children can look at and enjoy. The images also reflect around the room which has a constantly changing light. The lamp when not in use is not very nice to look at, but when it is lit in the dark it literally comes to life.

Safety
There is no sharp bits. However the bulb is easy to get to for very young children. The bulb is hot and they could burn them selves on it. There is a corded switch which makes it easier and safer to switch the lamp on and off. The lamp sits on four wooden feet which makes it fairly stable.

Size
Width: 160mm
Height:195mm

Target Market
Aimed at young children, can be used as a night light so they are not scared of the dark

These photographs show the small metal dome where the cone shaped point upon which the lightweight inner cylindrical shade spins on.

At the lamp have spherical turned wood to finish off the tops of the legs. They have pushed on to little points that are part of the metal structure. This photograph also shows how the outer layer has been glued round the frame.

This photograph shows how the wire frame has been welded to a metal disc that the light bulb holder is attached to. The metal has been powder coated so that it does not rust.

Corded Switch makes it easier and safer to turn the lamp on and off.

This photograph shows the underside of the lamp. It shows how the metal frame has been joined together. It also shows that the wooden legs glued to the frame.

A *This student undertook product analysis of several similar products which helped him to write a detailed list of design criteria for his own product*

Café Que and Access FM

These initials might be useful to you when analysing existing products. They simply provide you with a list of prompts for areas that you might comment on. They can also be used as a way of comparing products against one another. For example, rating each product 1 to 5 against each category would give you some numerical data that could be used for comparisons.

Cost
Appearance
Function
Ergonomics

Quality
User
Environment

Aesthetics
Customer
Cost
Ergonomics
Size
Safety

Function
Materials

Activity

Analyse a range of different commercial products throughout the course and keep a folder of these. They might save you a great deal of time when you undertake your controlled assessment.

Learning outcomes

By reading these pages it is hoped that you will be able to undertake your own product analysis in enough depth to be able to move your project forward quickly. This should help you to write design criteria for your own product.

13 Development and design proposals

13.1 Imaginative ideas

Initial design criteria

Having undertaken your initial research, you should put together some outline design criteria. These are the main considerations you need to be aware of when designing and will probably be based on the list of areas you used when undertaking your product analysis.

At this stage, it is likely that much of your design criteria will be based on what the product needs to look like. Basing your ideas on what already exists and producing near copies is unlikely to gain you high marks. You need other strategies to overcome the problem of staring at a blank piece of paper or computer screen.

Objective

Explore a range of designing strategies appropriate to your product, and get your designing off to a good start.

Using mood boards, image boards and sample boards

These are commonly used by designers as a way of talking to clients about the overall look or feel of the proposed product. GCSE students often create mood boards as a starting point for generating ideas. The best examples are going to be those that contain a range of related images on a theme, rather than a collection of existing products. Sample boards are commonly used by fashion and interior designers to explore the colours and textures of materials. Sample boards are usually better when scanned or colour copied for inclusion in your folder.

Using nature as a starting point

Natural pattern and form is often used as a starting point for fresh ideas, especially when generating ideas for ceramics, jewellery and fashion items.

Using architecture as a starting point

Architecture can provide a great starting point when generating ideas for structural products such as lighting or furniture. Repeated elements (or patterns) are common features in architecture.

A Monisha has used the theme of 'Circus' as a starting point for generating new and exciting ideas for jewellery

B Sarah has taken inspiration from the patterns found on animals to develop a range of textured jewellery

Choosing an appropriate drawing media

The choice of drawing media can have a dramatic effect on the ideas you generate. Look at the examples in Photo C and try to work out what the students have used.

Try some of the following techniques.

Drawing with scissors
Cut out shapes and paste then on your sheet. This is particularly good for modular designs, such as storage, jewellery and decorative elements, and for symmetrical shapes such as turned lamp bases or crockery.

White crayon on black paper
This is particularly good for metal objects such as jewellery or lighting.

Pastels, etc.
Spread pastel dust onto the paper in a streaky manner – this is great for making drawings look like plastic. Use coloured stickers for buttons and add tone using coloured pencils. This is really useful for electronic products such as radios and telephones.

Ballpoint pen
As you cannot rub out the lines, it makes you think more carefully about the marks you make and the drawings tend to look lively.

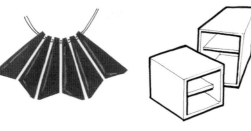

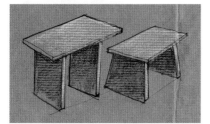

C *A range of drawing media. Try to work out how each has been produced*

Activity

Fashion design students often use **underlays** of figures to help them start their design drawings. They simply lay a translucent paper over the top and draw the clothing and the parts of the figure they want by tracing. This will work for all areas of product design. Trace some outline shapes of products and build up your own underlay collections. Use a photocopier to vary the sizes. Items such as bags, hats, electrical products, furniture and cushions can all be used in this way.

Commentary

The moderator likes to see that you have experimented with a variety of drawing media and different design strategies.

Do not worry if the sheets are different, or the paper or drawing style does not always match other sheets in your folder. For example, it is okay to paste in a drawing you did on lined paper in another lesson. It's how our mind works and your design folder should show a story of how your mind worked when generating ideas.

Learning outcomes

By reading these pages it is hoped that you will be able to develop your own method of showing how you have generated your first ideas, demonstrating originality, creativity and flair.

13.3 Considering wider issues

Taking account of a variety of social, moral, environmental and sustainability issues is an important part of your designing and these wider issues need to be considered as you are designing, not as a bolt-on at the end of the project. Ideally, the annotation that goes alongside your designing should highlight these issues.

Objectives

Understand the importance of the wider implications of your designing and making.

Explore ways of demonstrating this understanding.

Social issues

Some products encourage positive **social issues**, such as healthy eating, encouraging family interaction and including the needs of the disabled. Think about the social issues associated with the ideas you are generating. Are they all positive? If not, can you make any changes?

A Hugo has decided to apply an unusual surface decoration to his ceramics to highlight political figures that have been inspirational. Products can be used to reinforce beliefs

Moral issues

Moral issues include issues such as safety for the consumers. This must be a priority to ensure that the product you design is as safe as you can make it. Chapter 1 deals with this in more detail.

You will also need to think about how and where the product will eventually be manufactured. Many people believe that workers are often exploited in other countries to give us cheap products in the UK. Is that always the case? Others would argue that without these jobs people in other countries would starve. What do you think? Fairtrade is one organisation that tries to ensure the right deal for all concerned parties.

B Check out what the Fairtrade logo stands for. Do you support the organisation's ideals?

⚭ links

To find out more about the Fairtrade Foundation, see www.fairtrade.org.uk and see 6.2.

⚭ links

For more information on organisations related to **environmental issues**, see:

www.fscus.org

www.rainforestfoundationuk.org

www.global-standard.org

Environmental and sustainability issues

Making the product you design as environmentally friendly as possible is important for everyone. The materials you choose will have some impact on the environment. Choosing timber from sustainable forests, using materials that can easily be recycled, and reducing the materials you use are all areas that need to be considered. Using local materials can reduce energy costs associated with transportation. Using recycled materials can help further.

C *Two organisations that work hard to raise awareness of the importance of using sustainable timbers*

Reducing energy conservation and pollution

Specifying components that reduce energy consumption would be sensible and will help to access the higher grades, even if you cannot use them in the prototype you manufacture. For example, LED lamps are a great deal more energy efficient than the normal incandescent lamps, or even the low-energy fluorescent lamps. Specifying organic cotton in your design will reduce the amount of pesticides and fertilizers used.

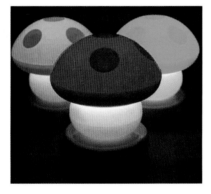

D *Choosing materials and components with **sustainability** in mind is important to access the higher grades. The lamp on the right contains an LED – how might this improve its sustainability?*

Activities

1. Make a list of all the things to avoid if you are going to make a sustainable product.

2. With your classmates, create a display of products that are beneficial or give pleasure to one group of users but cause offence or displeasure to another group. Refer to this resource when annotating your ideas.

Commentary

These wider issues must be made relevant to your specific product. Do not simply copy general material on moral, social and sustainability issues.

Consider using a code (this could simply be coloured dots) to draw attention to annotation made about these issues.

Learning outcomes

By reading these pages, it is hoped that you will understand how to provide the necessary evidence to show you have covered some of the wider implications of your designing and making.

13.4 Developing and modelling ideas

This is the stage of the designing in which you will need to resolve many of the design issues and start to finalise your thinking. The moderator will expect to see that you have considered alternative details, such as materials and constructions, and have undertaken any additional research needed. For a **soft model**, use cheap materials to quickly test your ideas. More accurate prototypes are needed to test specific features of your designs.

Soft modelling techniques

Simple modelling techniques using card and expanded polystyrene foam allow you to quickly test ideas and give you the impression of what the product would look like in 3D. It also gives you an opportunity to experiment with, for example, sizes and proportions, constructions, where controls might be housed and how mechanisms might work.

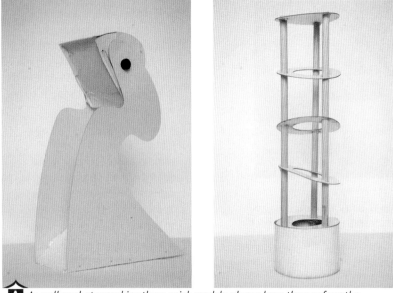

A As well as photographing these quick models, always keep them safe as the moderator might need to see them

Key terms

Soft model: a quick-to-make model using materials such as card, foam, fabric and wire.

Mock-up: a large-scale model (usually full size) that you can use to test main structural features.

Toile: a full-size mock-up garment usually made in calico.

Activity

Choose an existing product and try making several quick models to reflect the key features. This will help you develop your modelling skills.

Finalising sizes

When finalising sizes it is usual to undertake some modelling. At this stage, you also have the opportunity to undertake genuine primary research. It is far better to show how you have arrived at final dimensions through your own investigations, rather than rely on data from the internet or books.

Sample-making techniques

It is a good idea to test how you are going to make your prototype at an early stage. This is sometimes known as making a working **mock-up**. For garment designs you make a mock-up using a cheap fabric, such as calico, which is known as a **toile**.

B There is a wide range of construction kits that can be used to test ideas

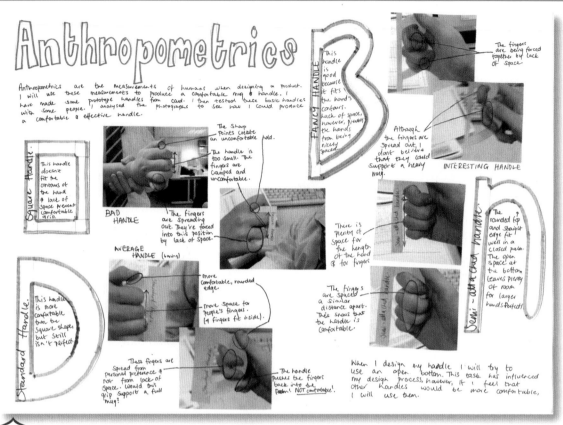

Anthropometrics

Anthropometrics are the measurements of humans when designing a product. I will use these measurements to produce a comfortable mug & handle. I have made some prototype handles from card. I then tested these basic handles with some people. I analysed the photographs to see how I could produce a comfortable & effective handle.

FANCY HANDLE
This handle is good because it fits the hand's contours. Lack of space, however, prevents the hands from being nicely spaced.

The fingers are being forced together by lack of space.

Although the fingers are spread out, I don't believe that they could support a heavy mug.

INTERESTING HANDLE

Square Handle.
This handle doesn't fit the contours of the hand & lack of space prevents comfortable grip.

The sharp points create an uncomfortable hold.

The handle is too small. The fingers are cramped and uncomfortable.

BAD HANDLE

The fingers are spreading out. They're forced into this position by lack of space.

AVERAGE HANDLE (swing)

More comfortable, rounded edge.

More space for people's fingers. (4 fingers fit inside).

Standard Handle.
This handle is more comfortable than the square shape but still isn't perfect.

These fingers are spread from personal preference & not from lack of space. Would this grip support a full mug?

The handle pushes the fingers back into the palm. NOT comfortable!

There is plenty of space for the length of the hand & for fingers.

The fingers are spaced a similar distance apart. This shows that the handle is comfortable.

Semi - just a curved handle
The rounded top and straight edge fit well in a closed palm. The open space at the bottom leaves plenty of room for larger hands-perfect!

When I design my handle I will try to use an open bottom. This task has influenced my design process, however, If I feel that other handles would be more comfortable, I will use them.

 Sam used modelling to test the shape and size of his mug handles

Using CAD

There are many computer programs that allow you to model your ideas.

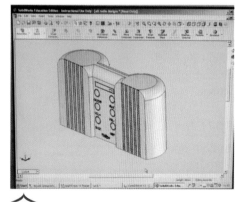 **D** *An example of computer-aided design*

Using CAM

Making models from cheaper materials using CAM is a sensible way forward. It will help you to test out your computer-aided drawings and give you a good idea as to what does and does not work well. Sometimes Styrofoam is used to produce 3D models, while card and MDF are often used for sheet materials. It is easy to make changes at this stage and build several models.

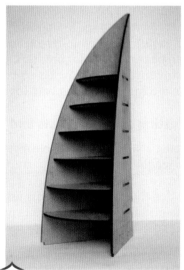 **E** *Testing out ideas using CAM might be very efficient in the available time*

Learning outcomes

By reading these pages it is hoped that you will understand the importance of modelling when developing your ideas.

 **kerboodle**

13.5 Presenting a design proposal

There is no single way of presenting a design proposal, although the following areas might provide you with a checklist of what to include:

- What the product will look like. This is often done by including a presentation drawing but could also be, for example, a model or mock-up.

- What size the product will be. This involves the all important dimensions, weights, volumes and values, etc. This will vary depending on whether you are working with materials or ingredients such as food, electronics or textiles.

- Materials that will be used and how they will need to be prepared for manufacture.

- How the product will be made.

Presentation drawings

The example in **A** and **B** show a variety of methods used for **presentation drawings**. In each case, it is clear what the product will look like. Drawings can be really helpful when presenting your proposal to your teacher or the moderator but do consider that it might be necessary to change some of the details once you start to make your prototype.

Working drawings

It is common to see formal **working drawings**, often as a result of designing using CAD. These can be very useful for products such as lighting or furniture but would not be expected, for example, from someone working with food or textiles.

Cutting lists, parts lists and patterns

Being able to select suitable materials and cut and shape them prior to manufacturing the product is often an important stage to record in your design folder. Cutting lists and parts lists will include all the materials and components needed. Textile products are usually manufactured using paper patterns.

Key terms

Presentation drawings: accurate visual representations showing, for example, colour, tone and texture, that are suitable to present to a client.

Working drawings: clear, detailed drawings showing sizes and construction details. They are often drawn to scale and based on British Standards Institute (BSI) conventions.

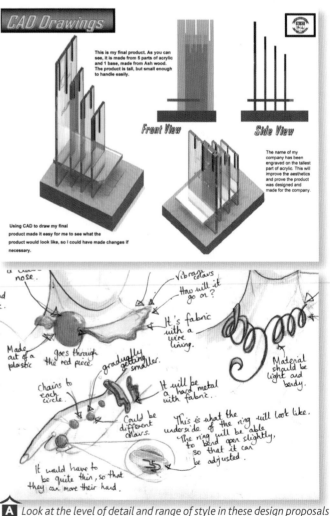

A *Look at the level of detail and range of style in these design proposals*

Activity

To practise the skills needed for presenting a good design proposal try some 'reverse engineering'. Take a simple commercial product that already exists and work out what information would have been needed to produce the product. Create the drawings and planning information and present this to your teacher.

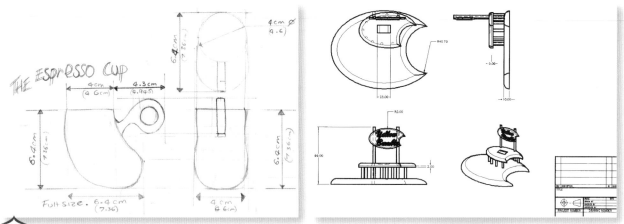

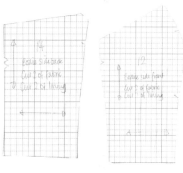

B *Orthographic projection is still one of the best ways of producing working drawings, whether hand drawn or CAD*

Ensuring that the most efficient use is made of the material and waste is minimised is also very important. This is called 'nesting' when using resistant materials and 'lay planning' when using textile materials.

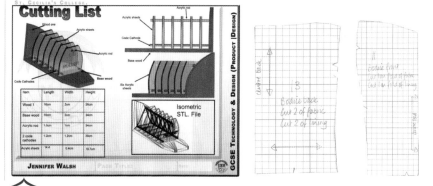

C *Ensuring that someone else can cut out the materials to the correct size from the information you supply is essential for the higher grades*

Manufacturing or product specifications

A product specification or **manufacturing specification** is one way of presenting, on a single sheet, a great deal of information that is essential for manufacturing your product. There are many alternative ways of doing this but as a general guide you should do the following:

- Write a concise description of your product.
- Illustrate your product with a clear drawing or photo of a model.
- Explain the scale of production.
- List the materials, providing a reference to a cutting list or patterns.
- List any pre-manufactured components in as much detail as you can.
- Give assembly/construction details.
- Provide the quality assurance procedures needed to ensure that each product is made to the same standard.
- Provide allowable tolerances, for example variations in weight or size that are acceptable.

13.6 Manufacturing decisions

You need to carefully plan your manufacturing, which includes deciding on the tools and equipment you will be using and how you will achieve the necessary levels of accuracy and finishing. Ideally, you should try to show off the making skills you have already developed rather than try to learn entirely new skills.

Selecting and choosing the materials

You should begin by listing the property requirements of the materials, such as whether they need to be waterproof, strong in tension or a good thermal insulator. This is usually the best time to undertake some research into materials, so that you can try to match the materials to the property requirements. Many choices will be based on **visual decisions**, such as colour and feel. Cost will often be a key factor as well as availability. It will be important to check with your teacher whether there are any **budgetary issues** or if materials will need to be ordered in advance.

Selecting and choosing the manufacturing methods

In much the same way as starting with material properties, it is useful to start by listing the components your prototype needs before working out the best way of making the prototype. This stage provides a real opportunity to experiment with actual materials and constructions. Making samples and mock-ups is a very useful task at this point, too. It is also an important time to consider alternative methods of manufacturing.

Which tools and equipment?

There are many ways of doing a similar job. Decisions on which tools and equipment to use will depend on:

- availability of equipment
- the numbers of each component needed
- the levels of accuracy needed
- whether you need to learn new skills.

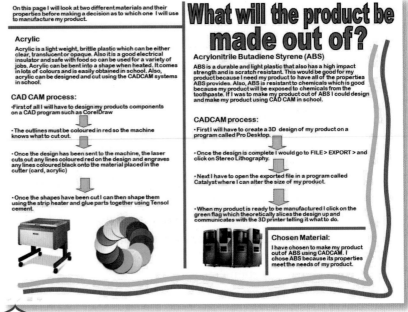

On this page I will look at two different materials and their properties before making a decision as to which one I will use to manufacture my product.

Acrylic

Acrylic is a light weight, brittle plastic which can be either clear, translucent or opaque. Also it is a good electrical insulator and safe with food so can be used for a variety of jobs. Acrylic can be bent into a shape when heated. It comes in lots of colours and is easily obtained in school. Also, acrylic can be designed and cut using the CADCAM systems in school.

CAD CAM process:

- First of all I will have to design my products components on a CAD program such as CorelDraw
- The outlines must be coloured in red so the machine knows what to cut out.
- Once the design has been sent to the machine, the laser cuts out any lines coloured red on the design and engraves any lines coloured black onto the material placed in the cutter (card, acrylic)
- Once the shapes have been cut I can then shape them using the strip heater and glue parts together using Tensol cement.

What will the product be made out of?

Acrylonitrile Butadiene Styrene (ABS)

ABS is a durable and light plastic that also has a high impact strength and is scratch resistant. This would be good for my product because I need my product to have all of the properties ABS provides. Also, ABS is resistant to chemicals which is good because my product will be exposed to chemicals from the toothpaste. If I was to make my product out of ABS I could design and make my product using CAD CAM in school.

CADCAM process:

- First I will have to create a 3D design of my product on a program called Pro Desktop.
- Once the design is complete I would go to FILE > EXPORT > and click on Stereo Lithography.
- Next I have to open the exported file in a program called Catalyst where I can alter the size of my product.
- When my product is ready to be manufactured I click on the green flag which theoretically slices the design up and communicates with the 3D printer telling it what to do.

Chosen Material:
I have chosen to make my product out of ABS using CADCAM. I chose ABS because its properties meet the needs of my product.

A James's choice of material and method of manufacture was influenced by the individual materials' properties

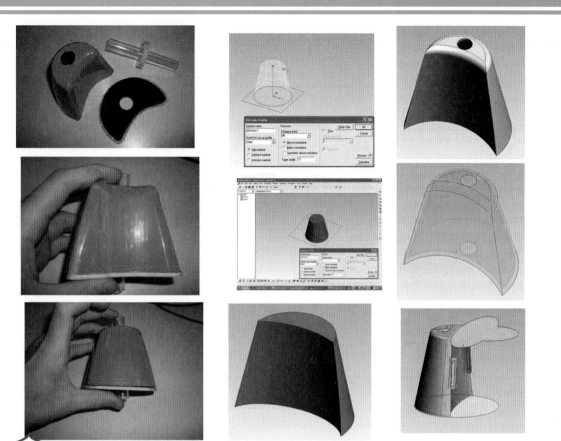

B Mark experimented with two different manufacturing methods before making his final product. Remember there is always more than one way of solving a manufacturing problem

C Choosing to use CAM is certainly a big advantage when several identical parts are needed

Learning outcomes

By reading these pages it is hoped that you will be able to make key decisions about how you will manufacture your prototype.

You will be expected to demonstrate a range of making skills throughout your Controlled Assessment task. These skills could include the work you do at the modelling stages, any testing you undertake, as well as the skills you demonstrate in the making of your final **prototype**. Ideally, there should be hand skills and computer-aided manufacturing skills and/or machine skills. Working with a range of different processes can make a big difference to your marks.

Making diaries

One of the best ways of keeping a record of the entire making you undertake is to keep a 'making diary'. This can provide a great deal of evidence, showing the moderator what manufacturing work you have done and how demanding it has been. Providing a simple photographic record, with a few added notes, is a good method of recording this work.

Objective

Understand what is expected in terms of demonstrating a **range** of making skills.

⊂⊃ **links**

For more on quality assurance, see 14.3.

⊂⊃ **links**

For more advice about showing how demanding your manufacturing work has been, see 14.2.

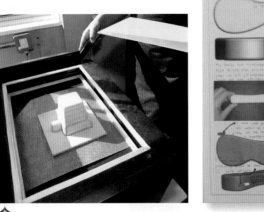
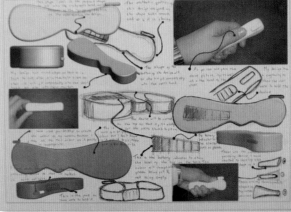

A *Photographing each process can be a simple way of evidencing the range of making skills you have demonstrated*

Using a variety of materials

You might have the opportunity to include a variety of materials in your prototype, although this is not always possible. Experimenting with a variety of materials at the development stages would be one way of ensuring you demonstrate a range of making skills.

Making manufacturing aids provides another opportunity to use a variety of materials.

Using a variety of processes

It is not always possible to include a range of different processes in your final prototype, but there may be opportunities to experiment with a range of alternatives at the development stages. This can be especially

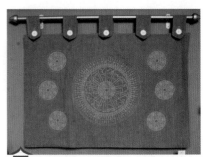

B *This student manufactured a wall hanging, cushion and rail, and also made a die-cutting tool to manufacture a swing label, thus demonstrating a wide range of skills*

useful in helping you finalise your materials and manufacturing processes. It can also provide ideal evidence that you have considered alternative methods and made informed choices.

When you have a design worked out in your head, try making part of it several times using different materials and processes. Make a list of all the different skills you have used each time.

Including packaging

Many students include packaging, labelling and user guides or assembly instructions as a way of extending their range of making skills. Others include **point-of-sale (POS)** displays and marketing materials to increase variety. This can be a great way of ensuring you have used a range of different making skills.

C *Experimenting with different techniques can make a big difference to the marks you gain for both development and making*

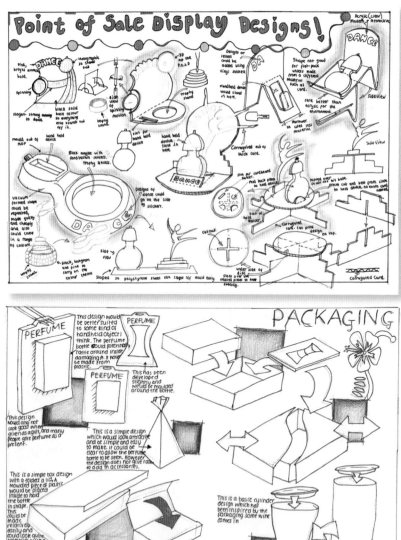

D *Hannah used a variety of making skills when presenting how her new perfume would be launched*

Key terms

Prototype: the first product made. Although often made using different materials and processes, this should be almost as good as the commercially made product and should allow you to fully test your ideas.

Point-of-sale (POS): where the product is displayed for sale. A POS is often part of a new product launch. It may be a unique display stand with key information about the product.

Commentary

Make sure you can demonstrate a range of making skills at the development stage, by experimenting with different materials and processes.

Learning outcomes

By reading these pages it is hoped that you will understand what is expected in terms of demonstrating a **range** of making skills.

Showing high level demand

When your work is assessed, one of the major considerations your teacher will have to make is to decide how demanding the work is. From a student viewpoint, understanding what is demanding is not easy and straightforward. On the previous pages we have seen that it can be useful to demonstrate a range of making skills. On these pages we will look at the degree of difficulty, but as a general guide, about 25 hours of making should be evidenced throughout both the development stages as well as during the making of the final prototype.

Working independently

To attract the highest marks you need to work **independently**. AQA expect all students to require some assistance and advice during their Controlled Assessment task but if you rely too heavily on help from your teacher or a technician then you will not be able to access the highest marks. This is particularly important when using CAM. The best way of showing what you have done is to use a mixture of screen dumps and photographs to simply show how you have prepared your file for cutting, how you have set up the machine and what you did independently of teacher or technician support.

Complexity

It is a common misunderstanding that your design has to be very **complex** to attract the higher grades. Making difficult wood joints just to gain extra marks is unnecessary, and may be less likely to gain credit for commercial viability. However, the moderator will be looking to see evidence of a broad range of making skills or a depth of skills in a single or narrow range of processes.

It is expected that it will take you a fair time to produce what you make, and will show off some very good making skills.

Objective

Understand what is expected in terms of levels of demand.

⊙⊙ links

For information on preparing your Summative Evaluation report, see 15.1.

Key terms

Independently: on your own! This means asking for as little help as possible.

Complex: complicated or difficult.

Activity

Working with your classmates, arrange a large selection of prototypes from your D & T department into a rank order based on how difficult each one would be to make. Make a decision where you think the C grade boundary would be and see if your teacher agrees with you.

A *All of the GCSE prototypes shown here were demanding to make. Some used CAM a great deal, while others used mainly hand processes*

By reading these pages it is hoped that you will understand what is expected in terms of levels of demand.

What does quality assurance mean as part of your Controlled Assessment task? You will need to show the moderator that you have undertaken the necessary checks at each stage of the construction of your prototype. These checks are practical skills and the moderator is expecting to see evidence in your prototype as well as in your design folder.

Checking accuracy

There are lots of checks you might make which would provide great evidence. These include:

- checking sizes, weights, volumes, etc.
- checking that the shapes are accurately cut
- checking that the parts fit together well.

You might present this evidence as part of your manufacturing specification, as part of a (making diary) or within your summative evaluation report. Explaining how these checks were undertaken and the results found would be an effective way forward.

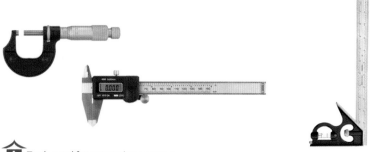

 Tools used for measuring accuracy

Checking for the quality of manufacturing

These checks, in commercial manufacturing, would relate to a very detailed manufacturing specification. Consumers will carry out such checks in a more **subjective** manner and may be concerned about:

- the overall finish of the product
- details such as joining material together
- the quality of the materials chosen
- the quality of standard components.

Now you know what people might be checking for, you might be able to establish procedures to ensure your product meets these requirements. Below are some areas you might consider.

Using CAM

Using computer-aided manufacturing techniques can be a great way of evidencing quality assurance. This can be especially effective if you are making multiple parts that are the same (assuming that the materials are the same and the **set-up** identical).

B Finishing details are very important for consumers but how do you ensure that the standard is always going to be the same?

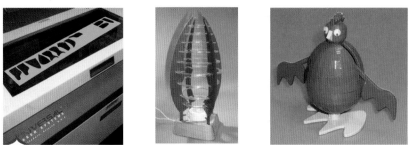

C *Using CAM can almost guarantee consistent quality in that particular part of the product*

Using manufacturing aids

Using manufacturing aids to help you to achieve accurate results can really make a big difference to the marks you achieve. This is especially useful if you make the manufacturing aid yourself. The purpose of manufacturing aids is to enable repeated tasks to be carried out the same every time. These might include:

- jigs for holding materials, for example when cutting or folding
- Fixtures for machinery, for example for drilling
- templates and patterns for accurate marking or cutting out
- formers for shaping materials accurately
- moulds for casting materials
- specialist tools such as die-cutting tools or measuring gauges.

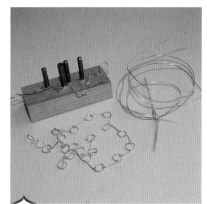

Once again, it is worth explaining why you used a particular manufacturing aid to improve the quality of the work.

Working to tolerance

D *Manufacturing aids to assist with quality assurance can easily be made in school*

You will have read in detail about what a manufacturing tolerance is. If you can detail what the + and − dimensions should be in your manufacturing specification, it will be easier to check your work. If you are working with food, you might describe the tolerances in weight or volume. You might include samples of the quality of joinings or surface finishes, for example. Remember, the smaller the tolerance, the higher the manufacturing costs are likely to be.

Commentary Make sure you know why CAM or a manufacturing aid would assure the quality is always the same for a particular operation.

Your manufacturing specification should include a statement about the tolerance limits.

links

For information on some of the advantages of using CAD/CAM, see page 81. You might list those relevant to your task in your design folder to explain why you have chosen the manufacturing method.

See 14.2 For how screen dumps and photos might be used to show your knowledge, see 14.2.

For more about your Summative Evaluation report, see 15.2.

links

Refer back to 5.4 for more information on quality assurance, quality control and manufacturing tolerance.

Activity

1

a Working as part of a team of three or four, look at a GCSE prototype from your D & T department and write a list of all the areas where you feel there are QA (quality assurance) issues. You might start by looking at all the areas where some QA decisions have clearly been taken. Then look at areas where there are clear weaknesses.

b When giving feedback to the rest of the class, decide on what could have been done to improve the overall quality.

Learning outcomes

By reading these pages it is hoped that you will understand the importance of quality assurance in your Controlled Assessment task and know what evidence you might provide.

Commercial viability

In GCSE Product Design, it is important that the prototype you develop has the potential for **commercial viability** and sale. This starts with identifying a realistic target market in the early stages, and ensuring that you continue considering this potential user throughout your designing.

Objective

Understand what you should look for when assessing commercial viability.

Considering costs

Many GCSE students keep a record of what it costs to build their prototype and try to make some estimates of the potential retail cost of their product. This is not a good way of considering costs and you are unlikely to be able to make realistic estimates in this way. It is much better to compare your prototype with similar commercial products and then make comparisons.

One of the first things to consider is the scale of the production. If you have been designing for a very broad market, it is perhaps feasible to consider large-scale production. If, however, you have been designing for a very narrow niche market then smaller batch production techniques are more likely. One mistake that GCSE students commonly make is to suggest that their product might be manufactured using injection moulding when only a small quantity are being made.

Key terms

Commercial viability: an assessment of how feasible your product will be to go into commercial production.

Styling: for example the colour, shape, form and pattern, which give a product its overall appearance.

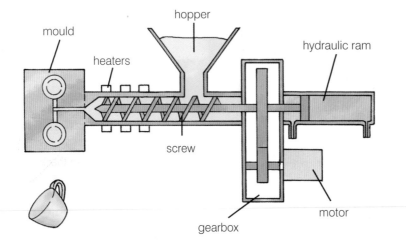

A *It is unlikely that injection moulding could be realistically considered for commercial production unless the volume of products is very high – typically 10,000 plus for aluminium moulds and 250,000 plus for steel moulds – as the cost of making the moulds would be too high*

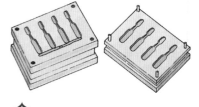

B *Sketch showing concept of multi-cavity moulds*

C *Injection-moulded product suitable for high-volume production*

Styling

Matching the **styling** of the product to the user is very important. Strong colour schemes and highly decorated products may be very attractive to some groups of people but might restrict sales. Take a look at some of the more successful products made in large volumes and you will see that they are often restricted in colour.

D *Remember to match your product styling to appeal to the consumer group you are aiming at*

Activity

1

a Take a small selection of products you feel are successful and write a list of the features you think contribute to the success.

b Now compare the list with the product you are planning to make and see how many of the same features you can incorporate into your product. What else do you need to do to make your prototype commercially viable?

Commentary

Making sure that your product looks just like it would in the shops will make a great deal of difference when considering this requirement. Ensuring that the styling matches the consumer group you have targeted and packaging the product appropriately will also help.

Remember

It is rare to use blister packaging for high value products.

Packaging

Packaging your product just as it would appear in the shops is a good way of meeting the requirement for commercial viability. Consider using an existing brand and presenting your product using that corporate style. This is often more convincing than trying to design your own logo and corporate identity. For a food product, for example, consider using a readymade carton and adding labels to explain the contents. This also works well for products such as lamps and small-scale furnishings. Many readymade cartons sold for storage solutions are particularly suitable for board games. You may need to think hard about how to include separate parts in your packaging, as well as how to keep the product secure inside the carton.

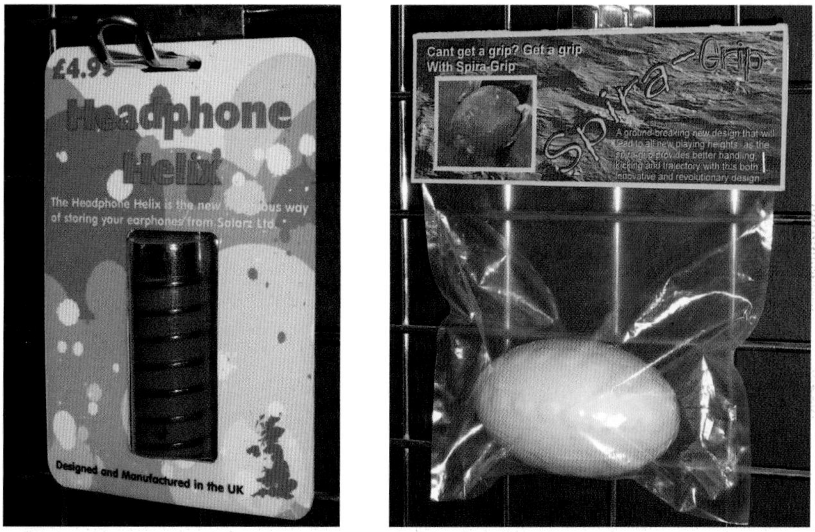

E *Producing effective packaging can help to make your prototype appear to be commercially viable*

Polythene bags are often used to package fabric products such as clothing and soft furnishings. Simple card inserts or fold-over hangers are common, as are swing labels. Blister packaging is easy to replicate and can provide a convincing packaging solution for certain products.

Learning outcomes

By reading these pages it is hoped that you will understand what is looked for when assessing commercial viability.

links

For more information on the functions of packaging and labelling, see 4.1.

To remind yourself about production scales and ways of organising production, see 7.1 and 7.2.

kerboodle

15.1 Testing and evaluating the prototype

Testing and evaluation should be continuous throughout your Controlled Assessment task. This is not an activity that should be left to the end. You are likely to have demonstrated your evaluation skills while:

- undertaking product analysis
- selecting the best ideas to develop further
- making and testing samples
- testing materials.

It is a good idea to plan to write a **summative evaluation** report. The emphasis in this should be on the prototype you have made and how successful it is, rather than on how well you worked and what problems you overcame. This is an important stage, even if you think that what you have produced is a failure or has lots of weaknesses. Don't forget to include several photos showing both the overall finished product and any constructional features.

Testing against the specification

It is important to test your product against the specification, whether that is the initial design specification, the manufacturing or the product specification. This is the easiest part of the evaluation as you can take each **design criterion** at a time and make judgements about how well your product meets it.

Comparing with existing products

Comparing your prototype with similar commercial products is very useful. These might be the products you analysed at the early stages of the Controlled Assessment task or they might be products your product would be competing against in the marketplace.

A Which features are better in the student-made lamp (on the right) and which are better in the commercially-made lamp?

> **Objective**
>
> Understand what the moderators expect from you in terms of testing and evaluating your prototype.

> **Key terms**
>
> **Summative evaluation:** a comprehensive report that summarises everything you have tested and evaluated throughout the task.
>
> **Design criterion:** a feature that must be included in your product. A number of these are called design criteria.

Chart for selecting best ideas

	Aesthetics	Cost	Customer	Environment	Safety	Size	Function	Materials	Total
Idea 1	3	2	3	4	5	3	4	3	27
Idea 2	2	3	3	4	4	4	5	4	29
Idea 3	1	1	2	2	3	3	3	3	18
Idea 4	4	4	4	4	5	4	4	4	33
Idea 5	5	5	5	5	4	5	4	4	37
Idea 6	5	4	4	4	5	4	4	3	33
Idea 7	3	3	4	3	3	4	5	5	30
Idea 8	2	2	1	3	3	3	3	4	21
Idea 9	1	1	1	1	4	3	3	3	17

B You may consider grading each criterion against a numerical scoring system or might simply explain it in words

Use your product specification as the basis for testing an existing commercially-made product. Use sketches or photos and notes to explain what improvements could be undertaken to the product to make it better-fit your specification.

Field testing

This is where you test your product in conditions that are as realistic as possible. For example, if you have made a toy for a five-year-old, it needs to be tested by a five-year-old (provided that the parent feels it is safe to do so). A bedroom lamp should be tested in a bedroom. Field testing takes time to arrange, so plan it in advance.

Third party evaluations

These are only valid if the people you ask are either experts or potential users. Asking your friends to comment on your work will not gain you credit. These evaluations also need planning and could perhaps be undertaken at the same time as your field testing. In the case of a toy, getting reactions from both the child and the parents may be useful.

Improvements

There is no such thing as the perfect product, so you will probably be able to see potential for further improvements. You could add notes to photos, or produce detailed drawings showing what improvements could take place.

C Independent testing is ideal if it can be arranged

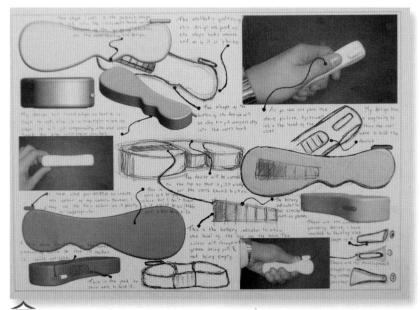

D Drawings of further design modifications might appear on your evaluation report

 Commentary Don't be a moaner! Even if you have faced a large number of problems throughout the Controlled Assessment task try to highlight the positive things you have achieved.

Do provide lots of photographic evidence and do try to get the prototype tested by someone who could be a potential user.

Learning outcomes

By reading these pages it is hoped that you will understand what is expected in testing and evaluating your prototype.

Modifications for commercial production

You are expected to manufacture a prototype closely resembling the final product that would be commercially produced. In many cases, the prototype will be made using the same techniques as commercial production. However, there will always be differences between making a single prototype and making a quantity, even if you are only planning for a small batch.

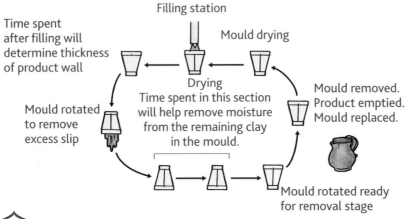

A *Slip casting used in school is almost identical to the commercial method shown here*

Filling station

Time spent after filling will determine thickness of product wall

Mould drying

Drying
Time spent in this section will help remove moisture from the remaining clay in the mould.

Mould rotated to remove excess slip

Mould removed. Product emptied. Mould replaced.

Mould rotated ready for removal stage

Objective

Understand what is necessary for explaining the changes from prototype to commercial production.

Commentary

You are not expected to produce a vast amount of work to meet this criterion. You should have been considering these issues throughout your task and it might simply be a case of summarising your decisions as part of your evaluation report.

links

Different production processes are covered in Chapter 9. See especially 9.1 and 9.2 for casting techniques, and 9.3 for moulding.

Key terms

Manufacturing plant: a factory or similar place where specialised machinery is arranged for efficient manufacturing. The 'plant' refers to the machinery and materials handling systems, for example.

Labour: refers to any form of work but is usually associated with people, so 'cheap labour' means that workers produce goods for very low wages.

Rapid prototyping: a CAM system that allows the prototype to be built up in layers directly from a CAD file. This allows the prototype to be tested before expensive injection moulds are made.

Scales of production

Make sure that you have explained how many items you think will be made. Will you use batch-production techniques, mass-production techniques or continuous-production techniques?

Manufacturing aids

If you are planning to use batch-production techniques, what manufacturing aids would be needed? For example, vacuum forming is a slow process, so many formings are done at the same time. Can you explain what this might look like? Individual holes are not marked out and drilled as you might have done in school. There would be a drilling fixture attached to the drill, or a multi-headed drill would be used to drill two or more holes at the same time. How might CAM be used?

B *A multi-headed drilling machine*

Manufacturing plant

Would you need highly specialised machinery and equipment or specialised tooling to be able to manufacture the product in quantity? You would if you were considering injection moulding, for example. Could any parts of your product be made elsewhere by specialist manufacturers? What might a production flow line be like? Would you expect the product to be assembled by hand or by machine?

C *Investment in the **manufacturing plant** is essential for effective, large volume manufacturing but remember that in many countries manufacturing is very dependent upon cheap **labour***

Materials

Have you considered the stock size of materials and whether anything could be done to maximise the materials and minimise waste? Are there likely to be any changes in the materials you used for making the prototype and commercial production?

Materials used for the prototype are often different at the commercial production stage. For example, vacuum forming high impact polystyrene sheet (HIPS) is often used at prototype stage to create parts that will eventually be manufactured using injection moulded ABS.

D *This prototype was made using a **rapid prototyping** system in a school but has been designed to be injection moulded*

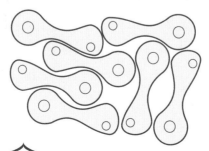

E *Nesting reduces materials wastage*

∞links

For information about nesting and lay planning, see 10.1. You might find it useful to revise these subject areas while thinking about your use of materials.

Learning outcomes

By reading these pages it is hoped that you will understand what is needed to explain the changes from prototype to commercial production.

16 Communication

16.1 Presenting your work

Paper or electronic portfolios?

There are many different methods you could consider to present your evidence of designing. AQA will accept sketchbooks, A4 or A3 paper **portfolios** or electronic submissions, either on CD or uploaded directly onto their website. Discuss the options with your teacher, who is the best person to advise you on the most appropriate method. Time on the Controlled Assessment task is tight, so whatever method you use must be efficient. You cannot afford to waste your time.

Handwritten or word processed?

As far as AQA is concerned, it does not matter which method you choose; many students choose to use a combination of handwritten notes alongside drawings and word-processed text for any extended writing, such as an evaluation report. Your decision will depend on access to computers, how neat your writing is and how well you can spell!

Hand drawn or computer-aided designing?

Again, AQA does not mind whether you use freehand drawing methods or CAD drawings throughout your designing. Many students include some of each. Generally, initial ideas are best done using freehand methods as they are generally quicker, whereas the accuracy and the ability to make changes can make CAD particularly effective at the development stages.

Objectives

Explore how best to present your work in order to match your skills and the type of task you have chosen.

Gain ideas and inspiration that will help you choose an appropriate presentation style.

Key terms

Portfolios: collections of some kind. A design portfolio is a collection of your design work.

Remember

You are not allowed to undertake large amounts of work at home, so you need to think carefully about the facilities available to you at school.

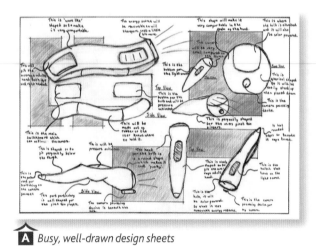
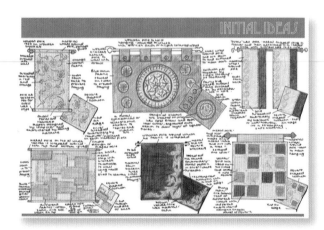

A Busy, well-drawn design sheets

Sketchbooks

These are often used at degree level and many designers use sketchbooks as one of their designing tools. However, these are usually combined with formal drawings. At GCSE level you need to be able to show a range of skills. Sketchbooks are particularly suitable for decorative products, such as jewellery, fashion and ceramics, when formal drawings are less likely to be needed.

Using flip-files

Whether A4 or A3, this method of working allows you to vary the sheets you include and make a good compromise between the formal and the sketchbook approach. Notes, drawings and photographs can be easily combined with formal technical drawings and word-processed reports.

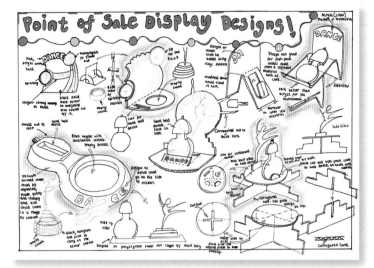

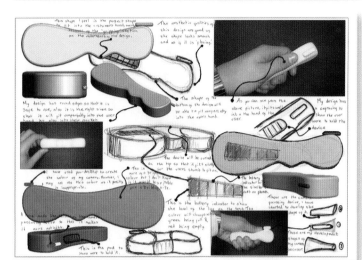

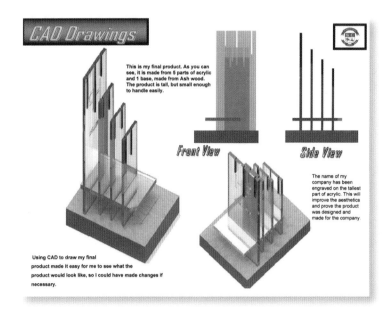

E-portfolios

There are several types of e-portfolio. The simplest is when your work is desktop published and saved to a CD rather than printed out. Freehand work can be scanned. To ensure that the moderator can open and access your work, AQA ask you to present this in a PowerPoint file or convert it to a PDF format. The second type of e-portfolio would be a multimedia presentation in which you might wish to include **sound bites** and video-clips, as well as drawings, notes and photographs. Again, PowerPoint is probably the best way of putting this together. AQA call each of these two approaches e-submissions. True electronic portfolios can involve submitting work from a variety of different formats.

AQA can provide your school with software that allows the work to be uploaded onto a secure part of the AQA website. The moderator will collect the work from there. This method of submitting your work is very new and few schools have experience of it – however, this facility does mean that it is possible to make use of the very latest technology.

Key terms

Sound bite: a very short piece of speech that can be recorded and included in a computer presentation.

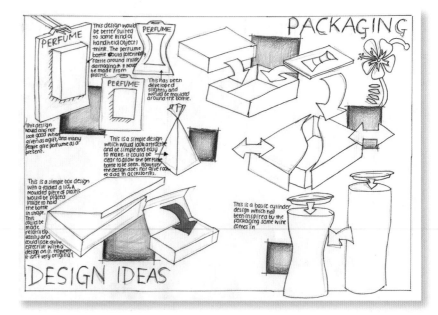

■ Presentation techniques

There is no such thing as the ideal presentation technique. Instead, it is a case of matching the style to the type of Controlled Assessment task you are undertaking. Most important is that you make your design portfolio concise and clear. You should be aiming at about 20–25 sheets of A3 paper with 12 pt writing (or a similar amount using other methods).

You will not gain the highest marks unless your work is concise and clear, so there is no point producing too much paperwork.

Activities

1 Try different styles of presentation during the first year of your course so that you can be confident of choosing the most appropriate route.

2 Try copying some of the drawings shown on these pages, using the same drawing styles and media. This will help you to develop your skills.

What sort of project should I choose for my Controlled Assessment?

AQA have produced a list of project outlines and it is likely that your teacher will have already reduced this list so that the choices available to you are well matched to the facilities in your school and what you have already done beforehand. You will be working on this controlled task for many weeks, so it is important that it is something that will hold your interest. Think carefully about the skills you already have and how you can show these off to AQA.

Can I produce a range of products?

It is possible to present a range of related products for assessment but these do all need to be part of the same task. For example, you might produce a range of jewellery pieces, or a product with packaging and a point-of-sale display stand. It is important that the range really demonstrates your best skills and does not just show what any KS3 student could achieve.

How much is my Controlled Assessment worth?

The Controlled Assessment is worth 60 per cent of the overall grade you are awarded for your GCSE in Design and Technology: Product Design. The rest of the grade is made up of the mark from the written examination paper. As the Assessment is worth more than the written examination, it does mean that it is an opportunity for you to gain the maximum amount of marks while you are still in the classroom and able to alter what you have done.

Commentary

Remember that you need to meet the assessment criteria, not just present your work well. Choose the method of presentation that is the most efficient in terms of time available and that matches your skills.

It is not expected that every page in a design folder will be presented to such a high standard. It is unlikely that you will have enough time to spend so much time on each sheet. The moderator is only looking for your skills once, you do not need to repeat them 20 times!

Plan your work before each lesson so that you can quickly present what you have done.

Who will assess my work?

Initially, your teacher will assess your work using the guidance provided by AQA. If there is more than one group taking Product Design, your teacher will need to compare their marking with the other teachers and agree a rank order for the work. This means that they will have to suggest a mark to AQA and all students with that mark should have done comparable work. This is a difficult task as no two teachers are always going to agree on how many marks your work deserves but they should be able to ensure that you get a higher grade if your work is better than your classmates'. AQA have a team of moderators. These are Design and Technology teachers from other schools who have been specially trained to make sure that your work matches that of students in another school who have been given a similar mark.

A *A concise design folder, sketchbook or e-portfolio is required*

Will the moderator see my work?

Not necessarily. The moderator will pick a sample of work from your school to compare with other schools'. Initially this will just be your design folder and the photos of what you have made. If they agree with the marks your school have given you, they will simply send the work back and let AQA know that your teacher's marking is in line with others. They may decide to visit and look at what you have made. It is possible that your mark may be adjusted after the visit. This is why it is vitally important that you provide you teacher with the evidence of everything you have done while working on this controlled task.

Will I get my work back from AQA?

In most cases the work is returned from the moderator very quickly but you are not allowed to take it home until October in case the school appeals against the grades that AQA have awarded. A handful of projects are chosen every year for training other teachers. It is a great honour if your work is chosen. In most cases it will be copied and returned before you are even aware it was sent away.

How big should my design folder be?

A concise folder is what is required. This is likely to be about 20 A3 sheets, which have normal size text and drawings, diagrams and photographs on them. Look at the examples on the following pages to see what is expected. If you are producing your folder electronically, there may be opportunities to include sound or video clips and so reduce the amount of writing you need to produce.

How much is my making worth?

Unlike coursework (for which is it was a fixed amount), it is not always going to be the same percentage. As a general guide, a Controlled Assessment is likely to be about 50 percent of the overall mark. You will gain marks for the making and modelling you undertake during the design and development stages, for any manufacturing aids you produce, as well as the final prototype. Of course, lots of designing can be undertaken using materials other than drawings, so there is lots of flexibility in the way you approach your Controlled Assessment task.

Learning outcomes

By reading these pages it is hoped that you will be able to negotiate with your teacher, and find the most appropriate way of presenting your work to match your skills and the type of task you have chosen.

Glossary

A

5th to the 95th percentile: the 'normal' range that product designers target.

Adhesive: a compound that bonds items together.

Aesthetics: The features in a product that make it visually appealing, such as colour, texture, shaping and styling features.

Alloy: a metal produced by combining two or more metals.

Analysis: discovering the important features of the design problem.

Anthropometrics: the study of human measurements.

Anthropomorphism: using human features on objects to improve the human interface.

Aseptic: sterile.

Assessment criteria: this is the list used by your teacher as a guide when marking your work.

Automation: the use of an automated production system.

B

Bakelite: a synthetic plastic named after its inventor – L.H. Baekeland.

Batch production: when a larger number of products are produced at the same time.

Biodegradable: break down naturally with the aid of rain and sunlight.

Blister packaging: packaging using a pre-formed plastic blister and a printed paperboard card which has a heat-seal coating.

Blobject: a product designed using CAD or CAM to reduce styling constraints.

Blow moulding: a manufacturing process for forming hollow plastic products.

Board: paper-based material, but thicker. Any form of paper weighing more than 200 gsm is classed as board.

Bonding alloy: the metal used in soldering and welding to form the join.

Brand development: creating and developing a strong product identity that will appeal to consumers.

Brand image: this identifies the company who made the product and gives a particular compression of its qualities.

Brik Pak: a term used to refer to a common form of aseptic packaging.

BSI: British Standards Institute.

Budgetary issues: issues involving the available finance.

C

CAD models: models designed digitally using special software.

CAD: computer-aided design.

CAM: computer-aided manufacture.

Casting: filling a space with liquid material until it becomes solid.

Casting pattern: the shape of the object required, usually made from timber and used to create the hollow shape in the sand.

CE: Conformité Européene

Cellulose: plant-based fibres used for paper-making.

Chiselling: a process used for chipping away pieces of timber, metal or concrete.

Client profile: a description of the lifestyle preferences of the chosen target market user.

CMYK: an abbreviation for cyan, magenta, yellow and black – the essential colours used in all colour printing.

CNC: computer numerical control.

CNC robots: machinery controlled by computer numerical control for use in manufacturing.

CNC routing: routing controlled by input from a computer.

Combined material: A material made from two of more different materials, with each material still retaining its own identity in the new material.

Commercial viability: an assessment of how feasible your product will be to go into commercial production.

Company logo and trademark: company symbols and service marks used to advertise and display products.

Complex: complicated or difficult.

Compliant materials: materials that are flexible, such as textiles and some plastics.

Components: basic electronic units connected to other units by being soldered onto a printed board to create an electronic circuit.

Composite materials: materials made from two or more different materials, with each material still retaining its own identity in the new material.

Compression moulding: moulding using heat and a two-part mould to squash the material into form.

Consultancy: an agency that provides professional advice.

Contemporary: belonging to the present day.

Continuous improvement: making designs better.

Continuous production: highly automated manufacture that runs continuously.

Corrosion: the deterioration of a metal, usually caused by a chemical reaction, and often linked to its environment.

Craftsmanship: specialised skills using tools in specialist areas.

Creasing: squashing the card so that it can easily be folded.

Creativity: the ability to create or to come up with new ideas.

D

Design considerations: areas that need to be considered when designing. These are often decided from analysing existing products and would largely be concerned with the user and areas, such as aesthetics, safety, usability and costs.

Design criterion: a feature that must be included in your product. A number of these are called design criteria.

Design engineers: People who try to solve design problems; they do not try to create new designs.

Design strategy: a method of generating or developing ideas.

Designer: a producer of designs which fulfil a need or fashion trend.

Die cutting: (general) a method of cutting and creasing material using a simple press knife principle.

Die cutting: a technique used in the printing process, involving cutting through with a blade attached to a former. (A former is used to repeatedly form a piece of material into a desired 2D or 3D shape.)

Digital music: analogue music is transferred into a computer data file.

Digital media: a form of advertising on the computer.

Dowels: circular sectioned pegs made from beech or other hardwoods.

Drape forming: a technique used for forming sheet materials.

Drilling: making cylindrical holes in solid materials using a rotary action.

E

Earthenware: a ceramic material fired at low temperatures.

Economies of scale: mass production allows products to be produced at a lower cost.

Electrical: operated using electricity.

Electrolysis: using electric currents to transfer particles from one item to the surface of another.

Electronic data interface (EDI): the transfer of structured data from one computer system to another without human intervention

Empirical design: a trial and error approach in design.

Environmental issues: we all rely on resources from the same planet. Everything we do impacts upon the planet in some small way. The trick is to improve the environment, not to damage it.

Ergonomics: the study of size, comfort and safety in relation to the human and the product.

Exclusive: excluding people by failing to meet their needs.

Extrusion: a technique involving the melting of raw plastic, which is then formed into a continuous profile.

F

Ferrous metal: a metal containing iron.

Filing: a pressing and dragging process to waste away materials.

Finite: limited.

Flair: a stylish quality in designs, which many people would want to own.

Flexible manufacturing: manufacturing controlled by a host computer that will log and sequentially operate several jobs.

Flux: a chemical that makes the bonding alloy flow more easily.

Food poisoning: an illness contracted by consuming contaminated food or drink.

Forging: a method of shaping metal using compressive forces.

Form: the structural beauty of a shape.

Fossil fuels: coal, oil and gas.

Fulcrum or pivot: the point on which a lever is balanced when a force is applied.

Function: how a product works.

Functionality: what a product does.

G

Galvanising: the process of coating steel with a thin layer of zinc to prevent corrosion.

Geometry: regular mathematical shapes, which are often found in design.

Gizmo: a small, multi-functional device.

Glaze: a way of sealing and decorating a ceramic product.

Greenhouse gases: gases, such as those produced by burning fossil fuels, which are linked to global warming.

H

Hardening: the heating of steel to 720 °C and cooling it rapidly in water to make it harder (also called quenching).

Hardwood: a natural material generally sourced from a deciduous tree.

Human interface: the relationship between the product and the user.

I

Inclusive: meeting everyone's needs.

Independently: on your own! This means asking for as little help as possible.

Injection moulding: a manufacturing process used for the production of plastic objects in large quantities.

Innovation: novel new ideas that usually display some very original thinking.

Intuitive design: design based on your past experiences.

J

Just-in-time (JIT) production: production organised so that the supply of materials is simultaneous with the need.

K

Kiln firing: a method of 'fixing' clay or ceramics by heating it to around 1 000 °C.

L

Labour: refers to any form of work but is usually associated with people, so 'cheap labour' means that workers produce goods for very low wages.

Laminate: a material formed by bonding two or more layers together; the layers are usually different materials.

Laminated: the process by which layers of material have been joined together.

Laminating: strengthening the material by building it up in layers with the same or another material.

Laser: amplification of an output of light producing an intense beam.

Laser cutters: tools for cutting, scoring or engraving; they use an infrared beam to laser out waste.

Line bending: the heating and bending of thermoplastic sheet material.

M

Manufactured boards: timber-based products produced by an industrial process involving gluing smaller pieces together.

Manufacturing considerations: areas that need to be considered when manufacturing. These are often decided by analysing existing products and would depend largely upon the scale of production required and the material properties.

Manufacturing plant: a factory or similar place where specialised machinery is arranged for efficient manufacturing. The 'plant' refers to the machinery and materials handling systems, for example.

Manufacturing specification: a collection of specific manufacturing information that would usually accompany a detailed working drawing.

Market: the target group a product is aimed at.

Market pull: how consumer demand leads to product development.

Mass production: manufacturing in high volume.

Mass-produced: made in great quantity by a standardised process.

Mechanical: working via a mechanism without direct human intervention.

Mock-up: a rough prototype made at low cost.

Modelling: a way of developing part or all of a 3D product using card, clay, foam, wood or CAD.

Moderator: this person will be a teacher from another school who has been trained by AQA to ensure that the marks your teacher awards are in line with AQA standards.

Moral issues: this is about being fair and honest. Making sure that the decisions you make do not harm others or make them feel uncomfortable. People with strong morals are thought to be honest and decent.

N

Natural fibres: fibres made from animal, plant or mineral sources.

Natural forms: shapes and images from nature.

Non-ferrous metal: a metal that does not contain iron.

Nutrients: substances from food that give us energy, help repair body tissues, and regulate body functions. There are six different types of nutrients: carbohydrates, fats, proteins, vitamins, minerals and water.

Nutrition: how our bodies take in, and use food.

Nutritional value: the amount of energy that a food gives you when you eat it. It is measured in either calories (kcal) or joules (kJ).

O

Obsolete: no longer working or useful.

Offset lithography: the name commonly used for offset lithography, which is the most common form of commercial printing.

One-off production: the making of a single, unique product.

Orthographic projection: drawings showing three dimensions in 2D.

P

Parison: a tube-like piece of plastic with a hole in one end, through which compressed air can pass.

Patterns: repeating shapes or objects.

Photochemical process: a process that uses chemicals and light to prepare metal plates for use in printing.

Physiological: relating to the body and its movement.

Piezoelectric: describes the ability of some materials to generate electricity when deformed by mechanical pressure. This effect is also reversible, causing piezoelectric materials to deform when a small voltage is applied.

Planing: shaving thin layers of timber from the surface.

Podcast: audio or video media file distributed over the internet – named from the words 'iPod' and 'broadcast'.

Point-of-sale (POS): where the product is displayed for sale. A POS is often part of a new product launch. It may be a unique display stand with key information about the product.

Polishing: a surface application applied to wood, metal and plastic.

Polymerisation: the process of creating a synthetic plastic.

Porcelain: a ceramic material fired at very high temperatures.